CHOCOLATE

...A Healthy New Image!

By Donna M. Ortman

CHOCOLATE . . . A Healthy New Image!
by
Donna M. Ortman

First Printing — June 1989

Copyright © 1989 by
Sweet Dreams Publishing
P.O. Box 3856
Regina, Saskatchewan, Canada
S4P 3R8

Canadian Cataloguing in Publication Data

Ortman, Donna Marie, 1960 —
    Chocolate . . . a healthy new image!

    Includes index
    ISBN: 0-919845-71-1

1. Cookery (Chocolate).   I. Title.
TX767.C5078   1989   641.6'374   C89-098063-2

Photography by:
Patricia Holdsworth
Patricia Holdsworth Photography
Regina, Saskatchewan

Illustrations by:
Mark Czarnecki
Calgary, Alberta

Dishes and Accessories Compliments of:
Junors The Kitchen Collection

Designed, Printed and Produced in Canada by:
Centax Books, a Division of M•C•Graphics Inc.
Publishing Director and Food Stylist: Margo Embury
1048 Fleury Street, Regina, Saskatchewan, Canada   S4N 4W8
(306) 359-3737 / 359-7580

♣ TABLE OF CONTENTS ♣

▲ ACKNOWLEDGEMENT ▲

The idea for this project originated through several ladies with whom I worked. They, like my family and friends, have teased me about my love for chocolate, but without them this book would never have been written. Special thanks to:

Marie Fitzpatrick

Kim Jakes

Deloise McKnight

Glenna Otto

Pauline Schmidt

Paulette Schumacher

Thanks, also, to my mom, Jean, and family members Linda, Irene and Bonnie for all of their testing help in the kitchen.

▲ HOW TO USE THESE RECIPES ▲

Whenever my recipes call for flour, I use all-purpose flour. When oil is called for, I use 100-percent vegetable oil. When whipped cream is used as a frosting, I flavor it with a bit of sugar and vanilla which is folded in after cream is partially whipped, 1 tbsp. (15 mL) sugar + 1/2 tsp. (2 mL) vanilla for every 1 cup (250 mL) of cream. I also find that several of the recipes taste better on the second or third day, after the ingredients have had a chance to blend and the flavors ripen. The recipe comments identify these recipes.

 FOREWORD

"CHOCOLATE . . . A Healthy New Image!" is a humorous attempt at putting chocolate in a more favorable light, in terms of nutrition. This book is put together purely for fun — as an answer to everyone who teases me about my lifelong love affair with chocolate. Designed as a spoof on Canada's Food Guide, these chocolate recipes all have ingredients that fall into the five basic food groups (milk, fruit, vegetables, bread/cereals and meat — with nut alternatives). The meat recipes are the most fun, some of them like "Turtle" Cake and Chocolate "Moose", have meat only in their titles. Now, when anyone asks, I can honestly say that I eat faithfully from the five basic food groups.

In keeping with a healthy new image for chocolate, health-conscious cartoons, engaged in all forms of physical activity, are incorporated throughout the book. I want to stress the fact that one can indulge a sweet tooth, yet still maintain an active and fit lifestyle. Chocolate facts, some historical, others scientific, emphasize that chocolate actually does have beneficial properties, from inhibiting tooth decay, to providing quick energy, to enhancing your love life.

"Chocolate . . . A Healthy New Image!" features thoroughly tested, no-muss, no-fuss recipes. They are simple to prepare, with easy-to-follow instructions. The end result is delicious and requires little time or effort. These recipes are perfect for everyday use and suggested sauces and garnishings give special-occasion glamor, when you want it.

This book is a must for all chocolate lovers who adore the flavor but simply don't have the time to prepare elaborate, time-consuming dishes. This cookbook contains a wide variety of recipes, with something for every taste and every occasion. I hope that you enjoy "Chocolate . . . A Healthy New Image!" as much as I enjoyed putting it together for you.

Donna Ortman

Donna Ortman

▲ CHOCOLATE TIPS ▲

1. Chocolate should be stored in a dark, dry, cool place. Milk chocolate will keep up to 6 months, plain chocolate up to 1 year and cocoa for 2 years or more.

2. If chocolate has a white film or "bloom" on the surface, it has likely been frozen or refrigerated, causing cocoa butter or sugar crystals to float to the surface. Storage in a too-warm environment will give the same appearance. When the chocolate is melted the film will disappear.

3. DON'T ADD WATER to chocolate when melting it. Water or overheating causes chocolate to tighten and harden. If this happens, add 1 tsp. (5 mL) of vegetable oil or shortening for each ounce (30 g) of chocolate. Beat mixture until chocolate smooths out. Use the same technique if chocolate hardens when alcohol is added.

4. Cocoa powder is made from the chocolate liquor that remains after the cocoa butter is removed. It has no sugar and very little cocoa butter. The chocolate flavor is very intense and the fat content is much lower than chocolate. Cocoa is the most economical form to use in cooking, for attaining a good chocolate flavor.

5. To substitute cocoa for chocolate in baking, a 1 ounce (30 g) square of unsweetened chocolate equals 3 tbsp. (45 mL) cocoa and 1 tbsp. (15 mL) butter or shortening. To increase cocoa in a cake recipe, increase the amount of fat also and very slightly decrease the amount of flour. To break down the starch cells in cocoa, blend it with enough boiling water to make a thick paste, or use it when the batter mixture is to be thoroughly cooked.

6. To substitute unsweetened chocolate for semisweet, add 1 tbsp. (15 mL) sugar to each ounce (30 g) of melted chocolate. Remember, 3 ounces (90 g) of unsweetened chocolate plus 3 tbsp. (45 mL) of sugar equals 4 ounces (120 g) of semisweet chocolate.

7. To make chocolate curls, use a vegetable peeler or the similar shape on a grater. Use a grater for fine or coarse grated chocolate.

8. To make chocolate squares, triangles or cut-outs, spread melted chocolate on waxed paper. Let set. Cut out squares or triangles with a knife and use cocktail or cookie cutters for fancy shapes.

9. To make chocolate leaves, brush melted chocolate on the bottom of unblemished rose leaves. Let set, chocolate side up, on waxed paper. When set, carefully peel away green leaves and discard.

MILK

 CHOCOLATE MALT

It's like drinking a fudgsicle!

½ cup	brown sugar	125 mL
¼ cup	cocoa	50 mL
2½ cups	milk	625 mL
1½ cups	vanilla ice cream, softened	375 mL

In a saucepan, combine brown sugar, cocoa and milk. Bring to a boil over medium heat, stirring constantly. Cook for another 5 minutes, stirring often. Cool mixture and then beat in ice cream. Pour into 2 tall glasses and serve immediately or chill until glasses are frosted. Serves 2.

 CHOCOLATE MIST

You'll never drink Brown Cows again!

1 cup	chocolate milk	250 mL
½ cup	Irish Cream liqueur	125 mL
2 cups	vanilla ice cream	500 mL
	semisweet chocolate, grated	

In a blender, combine milk, liqueur and ice cream. Blend on HIGH until smooth and creamy. Pour into 2 chilled glasses and top with grated chocolate. Serves 2.

See photograph, page 35.

 CHOCOLATE SODA

The soda water gives this drink a refreshing zing!

¼ cup	cocoa	50 mL
2 tbsp.	corn syrup	30 mL
2	eggs	2
2½ cups	creamilk	625 mL
2 cups	vanilla ice cream (4 scoops)	500 mL
2 cups	chocolate ice cream (4 scoops)	500 mL
10 oz.	bottle chilled soda water	300 mL

In a blender, combine cocoa, corn syrup, eggs, creamilk and vanilla ice cream. Blend on HIGH until frothy, about 1 minute. Divide mixture into 4 large chilled glasses then add 1 scoop of chocolate ice cream to each glass and top with soda water. Serve immediately. Serves 4.

BEST-EVER HOT FUDGE SAUCE

The name says it all! It can be reheated again and again.... if it lasts that long.

1½ cups	corn syrup	375 mL
¾ cup	cocoa	175 mL
¼ tsp.	salt	1 mL
⅔ cup	butter OR margarine	150 mL
1 tsp.	vanilla	5 mL
	ice cream, any flavor	

In a saucepan, combine corn syrup, cocoa and salt. Stir constantly over medium heat until mixture begins to boil. Reduce heat and simmer for 3 minutes. Remove pan from heat and stir in butter and vanilla. Serve as a topping on your favorite ice cream. Makes 2 cups (500 mL) of sauce.

YOGURT CHOCOLATE MUFFINS

Nutritious and delicious!

¾ cup	brown sugar	175 mL
½ cup	melted butter OR margarine	125 mL
1 cup	unflavored yogurt	250 mL
1	egg	1
1½ tsp.	almond extract	7 mL
1¾ cups	flour	425 mL
⅓ cup	cocoa	75 mL
½ tsp.	salt	2 mL
1½ tsp.	baking powder	7 mL
1 cup	chocolate chips	250 mL

Combine brown sugar and butter. Add yogurt, egg and almond extract, stirring mixture until thoroughly blended. In a larger bowl, combine flour, cocoa, salt and baking powder. Add yogurt mixture to dry ingredients, stirring only until mixture is slightly moistened. Fold in chocolate chips. Spoon batter into greased muffin tins and bake at 350°F (180°C) for 25 minutes. Makes 12 large muffins.

 The word cacao comes from the Mayan Indian word meaning "bitter juice."

CREAM CHEESE BROWNIES

These look pretty and they taste pretty wonderful!

3 tbsp.	butter OR margarine	45 mL
½ cup	chocolate chips	125 mL
2	eggs	2
¾ cup	sugar	175 mL
1 tsp.	vanilla	5 mL
½ cup	flour	125 mL
½ tsp.	baking powder	2 mL
½ tsp.	salt	2 mL
½ cup	chopped nuts	125 mL
2 tbsp.	butter OR margarine	30 mL
½ cup	cream cheese, softened	125 mL
¼ cup	sugar	50 mL
1	egg	1
1 tbsp.	flour	15 mL
1 tsp.	vanilla	5 mL

Melt together 3 tbsp. (45 mL) butter and chocolate chips. In a separate bowl, beat 2 eggs, then add ¾ cup (175 mL) sugar and 1 tsp. (5 mL) vanilla. Mix until well blended. Stir in flour, baking powder and salt; add the chopped nuts. In another bowl, combine remaining ingredients and beat until light and creamy. Spread chocolate mixture into greased 9" (23 cm) square pan. Add cream cheese mixture and swirl until batter is marbled. Bake at 350°F (180°C) for 40 to 45 minutes. Frost with Chocolate Buttercream Frosting, recipe follows.

CHOCOLATE BUTTERCREAM FROSTING

1½ cup	icing (confectioners') sugar	375 mL
⅓ cup	cocoa	75 mL
2-3 tbsp.	milk	30-45 mL
1 tsp.	vanilla	5 mL
⅓ cup	butter OR margarine, softened	75 mL

In a bowl, sift together icing sugar and cocoa. Add 2 tbsp. (10 mL) milk and vanilla and mix well. Stir in butter and beat until frosting is smooth and creamy. Add remaining milk if a thinner frosting is desired. This will serve as a topping for a 9" (23 cm) square cake.

CHOCOLATE CREAM CHEESE COOKIES

These melt in your mouth!

⅔ cup	butter OR margarine	150 mL
⅓ cup	cream cheese	75 mL
⅔ cup	sugar	150 mL
1 tsp.	vanilla	5 mL
1½ cups	flour	375 mL
⅓ cup	cocoa	75 mL
¼ tsp.	salt	1 mL
24	pecan halves	24

Cream butter, cream cheese and sugar until fluffy. Add vanilla and beat well. In a separate bowl, combine flour, cocoa and salt. Blend into creamed mixture. Pat dough into a ball and chill overnight. Roll out dough to ¼'' (1 cm) thickness. Cut out shapes with cookie cutter and place on ungreased baking sheet. Press a pecan half lightly into each cookie. Bake at 350°F (180°C) for 8 to 10 minutes. Makes 24 cookies.

 The Aztec Indians revered cacao beans as a gift from paradise, sown on earth by a prophet to offer universal wisdom to those who ate them.

BUTTERMILK CHOCOLATE CAKE

So moist and so good!

2 cups	flour	500 mL
½ cup	cocoa	125 mL
2 tsp.	baking soda	10 mL
½ tsp.	salt	2 mL
1 cup	butter OR margarine	250 mL
1½ cups	sugar	375 mL
2	eggs	2
2 cups	buttermilk	500 mL
1 tsp.	vanilla	5 mL

In a small bowl, combine flour, cocoa, baking soda and salt. In a larger bowl, beat butter and sugar until fluffy. Stir in eggs, buttermilk and vanilla. Add dry ingredients to creamed mixture, stirring just to moisten batter. Pour batter into greased 9" x 13" (23 cm x 33 cm) pan. Bake at 350°F (180°C) for 40 minutes. Frost with Chocolate Buttercream Frosting, page 11.

 Cacao beans were so precious to the Aztecs that they used them as currency. 100 such beans would buy a slave.

EGGNOG CHOCOLATE CAKE

Truly "Egg"squisite!

1 cup	sugar	250 mL
½ cup	oil	125 mL
1½ cups	eggnog	375 mL
1 tsp.	vanilla	5 mL
2 cups	flour	500 mL
½ tsp.	salt	2 mL
2 tsp.	baking soda	10 mL
4 tbsp.	cocoa	60 mL

In a large bowl, combine sugar, oil, eggnog and vanilla. Stir in flour, salt, baking soda and cocoa. Blend mixture thoroughly. Pour batter into greased 8" (20 cm) square pan and bake at 350°F (180°C) for 40 to 45 minutes. Frost with Mocha Frosting, page 107.

 The Aztecs used cacao beans, combined with water, vanilla and red peppers, to prepare a royal drink called "xocoatl" or "chocolatl" meaning "sour water" since the drink was so bitter.

SOUR CREAM CHOCOLATE CAKE

Easy to make. Easier to eat.

½ cup	butter OR margarine	125 mL
2 cups	brown sugar	500 mL
2	eggs	2
1 cup	sour cream	250 mL
½ tsp.	vanilla	2 mL
½ cup	cocoa	125 mL
1 tsp.	baking soda	5 mL
1 cup	boiling water	250 mL
2 cups	flour	500 mL
¼ tsp.	salt	1 mL
1 tsp.	baking powder	5 mL

Melt butter; add brown sugar, eggs, sour cream and vanilla. In a separate bowl, dissolve cocoa and baking soda in boiling water. Blend cocoa mixture into first mixture. Stir in flour, salt and baking powder. Pour batter into greased 9" x 13" (23 cm x 33 cm) pan and bake at 350°F (180°C) for 35 to 40 minutes. Frost with Chocolate Buttercream Frosting, page 11 or Fudge Frosting, page 68.

 Chocolatl was a liquid so precious that Emperor Montezuma II, ruler of the Aztecs, served the drink in golden goblets that were thrown away after one use. Thus, chocolatl was considered more precious than gold.

CHOCOLATE CHEESECAKE

A very rich, very elegant dessert!

1½ cups	graham cracker crumbs	375 mL
⅓ cup	butter OR margarine	75 mL
1 tbsp.	unflavored gelatin (7 g envelope)	7 g
⅔ cup	water	150 mL
16 oz.	cream cheese, softened	500 g
1 cup	chocolate chips, melted	250 mL
10 oz.	can sweetened condensed milk	300 mL
1 tsp.	vanilla	5 mL
1 cup	whipping cream, whipped	250 mL
	toasted almonds, slivered	

Combine crumbs and butter. Press firmly into a 9″ (23 cm) spring-form pan. Chill. In a small saucepan, soften gelatin in water. Over low heat, dissolve gelatin completely. In a separate bowl, beat cream cheese and chocolate until fluffy. Add milk and vanilla. Beat until smooth. Stir in gelatin mixture. Fold in whipped cream. Pour into prepared pan. Chill for at least 3 hours. Before serving, garnish with almonds.

Variation: Garnish with fresh raspberries and whipped cream, or pour a pool of Raspberry Sauce, recipe follows, on individual plates and place a wedge of cheesecake on sauce.

See photograph opposite.

RASPBERRY SAUCE

Wonderful with any dense chocolate dessert, also great on chocolate ice cream.

2 x 15 oz.	pkg. frozen unsweetened raspberries, thawed	2 x 425 g
⅓ cup	sugar OR to taste	75 mL
3 tbsp.	orange liqueur (optional)	45 mL

Drain raspberries. Add sugar and liqueur and purée in food processor. Strain out seeds with a fine sieve, if you wish. Add juice to make sauce desired consistency.

Chocolate Cheesecake, page 16.
Raspberry Sauce, page 16.

MOCHA CHEESECAKE

This is like heaven on earth!

1½ cups	crushed chocolate wafers	375 mL
⅓ cup	melted butter OR margarine	75 mL
8 oz.	cream cheese, softened	250 g
10 oz.	can sweetened, condensed milk	300 mL
⅓ cup	cocoa	75 mL
1 tbsp.	instant coffee granules	15 mL
½ cup	hot water	125 mL
1 cup	whipping cream, whipped	250 mL

Combine wafers and butter. Mix well and press mixture into greased 9'' (23 cm) springform pan. With an electric mixer, beat cream cheese until fluffy; add condensed milk. In a small bowl, dissolve cocoa and coffee in water. Add to cream cheese mixture. Fold in whipped cream. Pour into crust and freeze until firm, about 4 hours. Serve with whipped cream, flavored with a coffee liqueur if desired.

 Montezuma drank chocolatl before visiting his harem because he considered it a fortifier.

CHOCOLATE ALMOND CHEESECAKE

A special-occasion spectacular.

1½ cups	crushed chocolate wafers	375 mL
⅓ cup	butter OR margarine	75 mL
16 oz.	cream cheese, softened	500 g
1 cup	sugar	250 mL
⅓ cup	cocoa	75 mL
2	eggs	2
5 tbsp.	almond liqueur	75 mL
½ cup	sliced almonds	125 mL
1 cup	whipping cream	250 mL
	toasted almonds for garnish	

Combine wafers and butter. Press into the bottom of a greased 8″ (20 cm) springform pan. Beat cream cheese, sugar and cocoa until well blended. Add eggs and 3 tbsp. (45 mL) of liqueur; beat until smooth. Pour mixture into crust and sprinkle with almonds. Bake at 375°F (190°C) for 35 minutes. Cool completely in pan before removing sides. Combine whipping cream with remaining liqueur and beat until stiff peaks form. Drop cream by spoonfuls or pipe cream around sides and in center of cake. Garnish with toasted almonds.

 Christopher Columbus first took cacao beans back to King Ferdinand of Spain in 1502 after his fourth trip to America. The beans were virtually ignored because no one knew what to do with them.

FUDGEY ICE CREAM CAKE

You can't get enough of this!

1½ cups	crushed chocolate wafers	375 mL
⅓ cup	melted butter OR margarine	75 mL
4 cups	vanilla ice cream, softened	1 L
1 cup	peanuts	250 mL
1 cup	chocolate chips	250 mL
⅔ cup	icing (confectioners') sugar	150 mL
½ cup	butter OR margarine	125 mL
1½ cups	evaporated milk	375 mL

Combine wafers and melted butter. Press firmly into a 9" x 13" (23 cm x 33 cm) pan and set in refrigerator to chill. When base is chilled, spread ice cream evenly over it. Sprinkle with peanuts, patting in gently. Place pan in freezer to set. In a saucepan, combine chocolate chips, icing sugar, butter and milk. Bring to a boil over medium heat then reduce heat and simmer mixture for 8 minutes, stirring constantly. Cool mixture then spread chocolate over ice cream. Store cake in freezer until ready to serve.

 Montezuma served chocolatl to Mexican conqueror Hernán Cortés and his army in 1519. The Spaniards were curious about the drink and wanted to see the cacao tree. They found the drink too bitter and decided to sweeten it with cane sugar.

ALMOND FUDGE ICE CREAM

You'll scream for more!

2	eggs	2
¾ cup	sugar	175 mL
2 cups	whipping cream	500 mL
1 cup	milk	250 mL
¼ cup	coffee liqueur	50 mL
¾ cup	chopped, toasted almonds	175 mL
8 oz.	jar hot fudge topping OR 1 cup (250 mL) of Best-Ever Hot Fudge Sauce, page 9	250 mL

Whisk eggs for 2 minutes. Gradually beat in sugar and whisk 1 more minute. Stir in cream, milk and liqueur. Pour into an ice cream maker and churn according to manufacturer's directions, approximately 30 minutes. Stir in almonds and fudge topping. Freeze overnight. Makes 1 quart (1 L).

See photograph on front cover.

CHIPPY PISTACHIO ICE CREAM

A nutty chocolate crunch!

4 cups	creamilk	1 L
¾ cup	sugar	175 mL
1 tbsp.	vanilla	15 mL
⅛ tsp.	salt	0.5 mL
1 cup	chopped chocolate chips	250 mL
1 cup	chopped, toasted pistachios	250 mL

Thoroughly combine creamilk, sugar, vanilla and salt. Pour into an ice cream maker and churn according to manufacturer's directions, approximately 30 minutes. Stir in chocolate chips and pistachios. Store in freezer. Makes 1½ quarts (1.5 L).

CHOCOLATE CHUNK ICE CREAM

Double chocolate deluxe!

½ cup	chocolate chips	125 mL
1¼ cups	milk	300 mL
3	egg yolks	3
⅓ cup	sugar	75 mL
1¼ cups	whipping cream, whipped	300 mL
3 oz.	semisweet chocolate *	90 mL

In a saucepan, combine chocolate chips and milk. Stir over low heat until chocolate melts. In a medium-sized bowl, combine egg yolks and sugar. Beat until thick and creamy. Gradually pour in chocolate mixture and continue beating. Return chocolate, egg mixture to the saucepan and stir constantly, over medium heat, until thickened. Remove from heat and chill until almost set. Fold in whipped cream. Freeze until partially set, about 30 minutes, and then beat again. Chop semisweet chocolate into ¼" (1 cm) chunks and stir into ice cream. Return mixture to freezer until very thick, about 30 minutes. Beat again and freeze until firm. Makes 3 cups (750 mL).

* Belgian, Dutch and Swiss chocolate are excellent choices. They may be purchased at chocolate specialty shops.

▲ *Cinnamon was added to chocolatl by the Spaniards and they decided it would taste better if served hot.*

# MOCHA ICE CREAM

M . . . M . . . M . . . marvellous mocha!

¼ cup	butter OR margarine	50 mL
1 tbsp.	instant coffee granules	15 mL
½ cup	brown sugar	125 mL
¼ cup	cocoa	50 mL
¼ cup	water	50 mL
1¾ cups	chilled evaporated milk	425 mL

In a medium-sized saucepan, combine butter, coffee, brown sugar, cocoa and water over low heat, stirring constantly until sugar dissolves. On medium heat, bring coffee mixture to a boil. Remove from heat and cool to room temperature. In a large bowl, beat milk with an electric mixer until thick and frothy. Pour into the chocolate mixture and beat until thoroughly blended. Freeze, uncovered, until mixture is slushy. Beat again and then refreeze until firm. Makes 2 cups (500 mL).

Hernán Cortés said, "A cup of this precious beverage permits a man to walk an entire day without food."

 CHOCOLATE CREAM PUFFS

So good that I call them Dream Puffs!

½ cup	water	125 mL
¼ cup	butter OR margarine	50 mL
¼ tsp.	salt	1 mL
1 cup	flour	250 mL
2	eggs	2

Filling:

1½ cups	whipping cream	375 mL
½ tsp.	vanilla	2 mL
5 tbsp.	cocoa	75 mL
1 cup	sugar	250 mL

To make puffs, combine water, butter and salt in a saucepan. Heat, over medium heat, until bubbles begin to form. Remove from heat, add flour and stir vigorously. Add eggs, 1 at a time, beating until mixture breaks from spoon when raised. Drop batter by tablespoonfuls onto greased baking sheet and bake at 450°F (230°C) for 30 minutes. Reduce heat to 350°F (180°C) and bake for another 15 minutes. Remove from baking sheet and let cool. To make filling, stir together cream and vanilla. Sift in cocoa and sugar. Whip mixture until stiff peaks form. Split open cooled puffs and fill with the chocolate filling.

Variation: Make Chocolate Sauce from Chocolate Baked Alaska Pie, page 26, or use your favorite chocolate sauce and drizzle sauce over filled cream puffs. Garnish with whole fresh strawberries.

CHOCOLATE BAKED ALASKA PIE

A blatantly decadent experience!

1¼ cups	crushed chocolate wafers	300 mL
⅓ cup	melted butter OR margarine	75 mL
¾ cup	chopped nuts	175 mL
¼ cup	brown sugar	50 mL
2 cups	chocolate ice cream, 4 scoops	500 mL
2 cups	vanilla ice cream, 4 scoops	500 mL

Chocolate Sauce:

½ cup	chocolate chips	125 mL
¼ cup	water	50 mL
¼ cup	sugar	50 mL
2 tbsp.	butter OR margarine	30 mL

Meringue:

4	egg whites	4
½ tsp.	cream of tartar	2 mL
½ cup	sugar	125 mL

To make crust, combine wafers, melted butter, nuts and brown sugar. Press mixture firmly into a 9" (23 cm) pie plate. Chill until firm. Scoop chocolate and vanilla ice cream, alternately, into crust. To make chocolate sauce, melt chocolate chips and water in a saucepan, over low heat, until smooth. Add sugar and boil for 2 minutes over medium heat, stirring constantly. Add butter and mix well. Drizzle half of chocolate sauce over ice cream. Freeze until very firm. To make meringue, beat egg whites and cream of tartar until foamy. Gradually add ½ cup (125 mL) sugar and beat until stiff peaks form. Cover ice cream completely with meringue. Bake at 450°F (230°C) for 3 minutes. Drizzle with remaining chocolate sauce. Serve immediately.

 MOCHA CREAM PIE

A creamy delight!

1½ cups	crushed chocolate wafers	375 mL
¼ cup	melted butter OR margarine	50 mL
2 cups	miniature marshmallows	500 mL
⅓ cup	milk	75 mL
2 cups	coffee-flavored ice cream *	500 mL
1 oz.	semisweet chocolate (1 square)	30 g
1 tbsp.	butter OR margarine	15 mL
	toasted, slivered almonds	

Combine chocolate wafers and melted butter. Press into a 9" (23 cm) pie plate. Chill. In a medium-sized saucepan, combine marshmallows and milk. Stir constantly, over low heat, until marshmallows are completely melted. Remove from heat and chill. Once chilled, add ice cream and mix well. Pour mixture into crust. Freeze several hours. When ready to serve, melt chocolate with butter and drizzle over the pie. Sprinkle with almonds.

* To make coffee-flavored ice cream, mix 2 tsp. (10 mL) of instant coffee granules with 1 tbsp (15 mL) of water and stir into 2 cups (500 mL) of vanilla ice cream.

 Hernán Cortés wrote to King Charles V, "Chocolatl is the divine drink which builds up resistance and fights fatigue."

CHOCOLATE CREAM CHEESE SOUFFLÉ

A fabulous make-ahead dessert!

1½ cups	sugar	375 mL
2 tbsp.	unflavored gelatin (2 x 7 g pkg.)	30 mL
2 cups	water	500 mL
2 oz.	unsweetened chocolate (2 squares)	55 g
4	eggs, separated	4
8 oz.	cream cheese, softened	250 g
½ tsp.	almond extract	3 mL
1½ cups	whipping cream, whipped	375 mL
	chocolate curls for garnish	

In a saucepan, combine 1 cup (250 mL) sugar with gelatin; add water, chocolate and egg yolks. Stir, over low heat, until gelatin dissolves and chocolate melts. Remove from heat and beat until well blended. Cool for 10 minutes. Slowly add chocolate mixture to cream cheese, mixing until blended. Add almond extract. Chill until slightly thickened. Beat egg whites until frothy. Gradually beat in remaining ½ cup (125 mL) sugar until stiff peaks form. Fold egg mixture into chocolate mixture. Fold in 1 cup (250 mL) of whipping cream, whipped. Wrap a 3″ (7 cm) collar of foil around top of a 6-cup (1.5 L) soufflé dish and secure. Pour soufflé mixture into dish and chill overnight. Remove collar and garnish with remaining ½ cup (125 mL) of whipping cream, whipped, and chocolate curls.

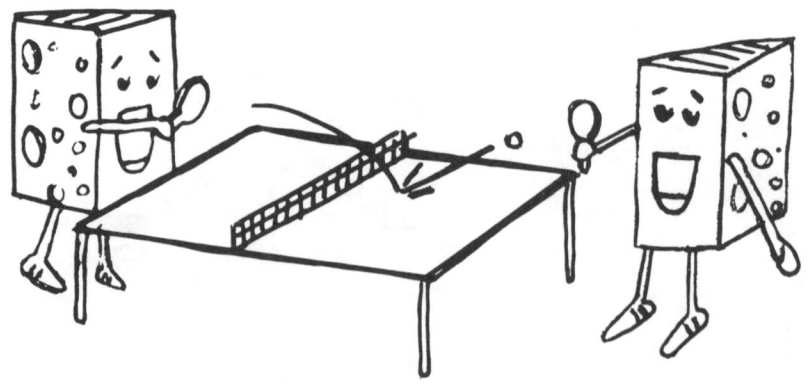

FRUIT

 APPLE NUT BREAD

Very good with your morning coffee!

2 cups	flour	500 mL
1¼ cups	sugar	300 mL
⅓ cup	cocoa	75 mL
1 tsp.	baking soda	5 mL
¼ tsp.	salt	1 mL
1 tsp.	cinnamon	5 mL
½ tsp.	nutmeg	2 mL
2	eggs	2
½ cup	milk	125 mL
¾ cup	melted butter OR margarine	175 mL
2 cups	peeled, diced apples	500 mL
1 cup	chopped walnuts	250 mL

Combine flour, sugar, cocoa, baking soda, salt, cinnamon and nut-meg. In a separate bowl, beat eggs then add milk and butter. Pour liquid mixture into the dry ingredients. Stir just until moistened. Fold in apples and walnuts. Spoon batter into a greased 5" x 9" (12 cm x 23 cm) loaf pan. Bake at 350°F (180°C) for 70 to 80 minutes. Makes 1 loaf. When cool, top with Milk Glaze, if desired. Recipe follows.

 MILK GLAZE

1½ cups	icing (confectioners') sugar	375 mL
2 tbsp.	hot milk	30 mL
1 tsp.	vanilla	5 mL

Combine all ingredients. Stir until sugar is dissolved.

 CHOCOLATE LEMON LOAF

Sensational flavor combination!

1 cup	butter OR margarine	250 mL
¾ cup	brown sugar	175 mL
½ cup	chocolate chips, melted	125 mL
3	eggs	3
1	lemon, grated rind of	1
2 tbsp.	lemon juice	30 mL
1 tsp.	vanilla	5 mL
2 cups	flour	500 mL
2 tsp.	baking powder	10 mL
6	crystallized lemon slices	6

Cream together butter and brown sugar. Mix in chocolate, eggs, lemon rind, lemon juice and vanilla. Add flour and baking powder, stirring until well blended. Spoon batter into greased 5″ x 9″ (12 cm x 23 cm) loaf pan and bake at 325°F (160°C) for 45 to 50 minutes. Frost with Chocolate Buttercream Frosting, page 11, in which 2 tbsp. (30 mL) of lemon juice have been substituted for milk and garnish with crystallized lemon slices. Makes 1 loaf.

Variation: To make a Lemon Glaze, for a more intense lemon flavor, combine 2 tbsp. (30 mL) of lemon juice with ¼ cup (50 mL) of white sugar. Pierce warm loaf all over with a skewer, pour lemon syrup over loaf and cool before removing from pan. Frost as above.

CHOCOLATE APPLE MUFFINS

A healthy treat for your lunch box.

½ cup	butter OR margarine	125 mL
1 cup	sugar	250 mL
2	eggs, beaten	2
1 tsp.	vanilla	5 mL
½ cup	coffee	125 mL
1½ cups	flour	375 mL
1 tsp.	baking soda	5 mL
½ tsp.	salt	2 mL
2 tbsp.	cocoa	30 mL
1 tsp.	cinnamon	5 mL
¾ cup	raisins	175 mL
¾ cup	chopped nuts	175 mL
1 cup	chopped apple	250 mL

Cream together butter and sugar. Add eggs, vanilla and coffee. In a separate bowl, combine flour, baking soda, salt, cocoa and cinnamon. Spoon creamed mixture into the dry ingredients, stirring just to moisten. Fold in raisins, nuts and apple. Spoon batter into greased muffin tins and bake at 350°F (180°C) for 35 minutes. Makes 12 large muffins.

 The Mayan mistress of Cortés assured him in 1519 that chocolatl was, indeed, a powerful aphrodisiac.

CHOCOLATE BANANA MUFFINS

These are great on the second day - the banana flavor is more pronounced.

2 cups	flour	500 mL
⅔ cup	sugar	150 mL
⅓ cup	cocoa	75 mL
2 tsp.	baking powder	10 mL
1 tsp.	salt	5 mL
1 tsp.	vanilla	5 mL
1	egg, beaten	1
1 cup	milk	250 mL
⅓ cup	oil	75 mL
1 cup	mashed banana	250 mL
1 cup	chopped nuts	250 mL

Stir together flour, sugar, cocoa, baking powder and salt. In a separate bowl, combine vanilla, egg, milk, oil, banana and nuts. Pour liquid ingredients into the dry mixture, stirring just to moisten. Spoon batter into greased muffin tins and bake at 350°F (180°C) for 20 to 25 minutes. Makes 12 large muffins.

 Because of chocolatl's reputation as being an aphrodisiac, it was forbidden to be consumed by Aztec women and 17th century monks.

MARASCHINO CHERRY MUFFINS

A Black Forest Muffin!

2 oz.	unsweetened chocolate, melted (2 squares)	55 g
⅓ cup	oil	75 mL
1	egg	1
½ cup	buttermilk	125 mL
½ cup	sour cream	125 mL
¾ cup	brown sugar	175 mL
¼ cup	maraschino cherry juice	50 mL
1 tsp.	vanilla	5 mL
2 cups	flour	500 mL
2 tsp.	baking powder	10 mL
½ tsp.	salt	2 mL
½ cup	chocolate chips	125 mL
½ cup	chopped maraschino cherries	125 mL

Combine melted chocolate and oil. Cool, then add egg, buttermilk, sour cream, brown sugar, cherry juice and vanilla. In a separate bowl, combine flour, baking powder and salt. Pour liquid ingredients into dry mixture, stirring just to moisten. Fold in chocolate chips and cherries. Spoon batter into greased muffin tins and bake at 350°F (180°C) for 25 minutes. Makes 12 large muffins.

Strawberry Chocolate Torte, page 41.
Chocolate Mist, page 8.

▲ CHOCOLATE FIG COOKIES ▲

Not too sweet, so they're easy on the "fig"-ure!

¾ cup	butter OR margarine	175 mL
¾ cup	sugar	175 mL
1	egg	1
2 tbsp.	milk	30 mL
1 tsp.	vanilla	5 mL
2 cups	flour	500 mL
2 tbsp.	cocoa	30 mL
2 tsp.	baking powder	10 mL
½ tsp.	salt	2 mL
2 cups	chopped, dried figs	500 mL

Cream together butter and sugar. Beat in egg, milk and vanilla. Add flour, cocoa, baking powder and salt. Mix well. Fold in figs. Drop batter by spoonfuls onto greased baking sheet. Bake at 350°F (180°C) for 12 to 15 minutes. Makes 48 cookies.

▲ *Hot chocolate became the drink of Spanish Aristocracy and the Spaniards managed to keep the art of the cacao industry a secret from the rest of Europe for nearly 100 years.*

ORANGE CHIP COOKIES

These cookies won't last long!

1 cup	butter OR margarine	250 mL
1 cup	brown sugar	250 mL
2	eggs	2
2 tsp.	vanilla	10 mL
2 tsp.	grated orange rind	10 mL
2 tbsp.	orange juice	30 mL
2½ cups	flour	625 mL
½ tsp.	baking soda	2 mL
¼ tsp.	salt	1 mL
1 cup	chopped walnuts	250 mL
1 cup	chocolate-orange chips	250 mL

Cream together butter and brown sugar. Add eggs, vanilla, orange rind and orange juice, mixing thoroughly. Stir in flour, baking soda and salt. Add walnuts and chocolate-orange chips. Mix well. Drop batter by spoonfuls onto greased baking sheet. Bake at 350°F (180°C) for 12 to 15 minutes. Makes 36 cookies.

 Around 1600, C. F. Ragionamenti wrote that in Mexico a chocolate drink was consumed that pleased and satisfied the body, giving it strength, nourishment and vigour in such a way that those accustomed to drinking it could not remain strong without it, even by eating other substantial things. He said that they appeared to diminish when they did not have such drink.

CHOCOLATE DATE SQUARES

These are very moist and they freeze well.

1 cup	chopped dates	250 ml
1 tsp.	baking soda	5 mL
1½ cups	boiling water	375 mL
½ cup	butter OR margarine	125 mL
1 cup	sugar	250 mL
2	eggs	2
1 tsp.	vanilla	5 mL
1½ cups	flour	375 mL
¾ tsp.	salt	3 mL
¾ tsp.	baking soda	3 mL
½ cup	brown sugar	125 mL
½ cup	chopped walnuts	125 mL
1 cup	chocolate chips	250 mL

In a large bowl, combine dates and 1 tsp. (5 mL) of baking soda. Pour boiling water over top and set aside to cool. In another bowl, cream together butter and sugar. Beat in eggs and vanilla. Blend into date mixture. Add flour, salt and ¾ tsp. (3 mL) baking soda and mix well. Spoon into greased 9" x 13" (23 cm x 33 cm) pan. In a small bowl, combine brown sugar, walnuts and chocolate chips. Sprinkle mixture on top of batter. Bake at 350°F (180°C) for 50 minutes.

See photograph on back cover.

 During the 1600's many Europeans believed that chocolate calmed fevers, cured chronic dyspepsia and prolonged life.

 APPLE CHIP CAKE

A moist and delicious cake!

1 cup	butter OR margarine, softened	250 mL
2 cups	sugar	500 mL
3	eggs	3
½ cup	water	125 mL
1 tbsp.	vanilla	15 mL
2½ cups	flour	625 mL
2 tbsp.	cocoa	30 mL
1 tsp.	baking soda	5 mL
1 tsp.	cinnamon	5 mL
2 cups	peeled, cored and finely chopped apples	500 mL
¾ cup	chocolate chips	175 mL

Cream butter and sugar; beat in eggs, then add water and vanilla. In another bowl, stir together flour, cocoa, baking soda and cinnamon. Beat flour mixture into the creamed mixture. Stir in apples and chocolate chips. Pour into greased and floured bundt pan. Bake at 325°F (160°C) for 75 minutes. Transfer to a rack to cool. Frost with Fudge Frosting, page 68.

STRAWBERRY CHOCOLATE TORTE

I never tire of this one!

2¼ cups	brown sugar	550 mL
¾ cup	butter OR margarine	175 mL
3	eggs	3
1½ tsp.	vanilla	7 mL
1½ cups	flour	375 mL
¾ cup	cocoa	175 mL
1½ tsp.	baking powder	7 mL
2 cups	whipping cream, whipped	500 mL
15 oz.	pkg. strawberries in light syrup, thawed	425 mL
	whole strawberries	
	chocolate shavings (optional)	

Cream together brown sugar and butter. Beat in eggs and vanilla. Combine flour, cocoa and baking powder; add to sugar mixture and mix thoroughly. Divide batter into 3 greased 8″ (20 cm) round pans and bake at 325°F (160°C) for 25 minutes. Remove from oven and let cool 10 minutes. Remove cakes from pans to cool completely. To assemble torte, spoon ½ of the strawberries and syrup on the bottom layer and top with ⅓ of the whipped cream. Place second layer over bottom layer and spoon on remaining strawberries and syrup and top with ⅓ of the whipped cream. Place third layer over second layer and top with remaining whipped cream. Garnish with whole strawberries or strawberry fans and chocolate shavings, if desired.

See photograph, page 33.

BLACK FOREST BUNDT CAKE

A personal favorite!

½ cup	butter OR margarine	125 mL
½ cup	sugar	125 mL
3	eggs	3
½ cup	grated almonds	125 mL
1 cup	grated chocolate chips	250 mL
1 tsp.	vanilla	5 mL
¾ cup	flour	175 mL
1 tsp.	baking powder	5 mL
½ tsp.	salt	2 mL
¼ cup	maraschino cherry juice	50 mL
1 cup	whipping cream, whipped	250 mL
	maraschino cherries for garnish	
	chocolate curls for garnish (optional)	

Cream together butter and sugar. Beat in eggs, 1 at a time. Stir in almonds, chocolate and vanilla. In a separate bowl, combine flour, baking powder and salt. Stir flour mixture into the creamed mixture. Spoon batter into greased bundt pan and bake at 350°F (180°C) for 35 minutes. Cool for 20 minutes and invert pan, removing cake from pan. Pierce top of cake all over with a fork and slowly pour the cherry juice over the cake. Let cake cool completely. Frost entire cake with whipped cream, and then garnish with maraschino cherries and chocolate curls, if desired.

See photograph, page 103.

▲ CHOCOLATE APRICOT CAKE ▲

Rich and tangy!

2	eggs, separated	2
1½ cups	sugar	375 mL
1¼ cups	flour	300 mL
½ cup	cocoa	125 mL
¾ tsp.	baking soda	3 mL
½ tsp.	salt	2 mL
½ cup	oil	125 mL
1 cup	milk	250 mL
1 tbsp.	vinegar	15 mL
2 cups	apricot jam	500 mL
1 cup	whipping cream, whipped	250 mL

Beat egg whites until foamy and then beat in ½ cup (125 mL) of sugar until stiff peaks form. In a separate bowl, combine remaining sugar, flour, cocoa, baking soda and salt. Stir in oil, milk, vinegar and egg yolks and beat until mixture is smooth. Gently fold egg whites into the batter. Pour mixture into 2 greased 9″ (23 cm) layer pans and bake at 350°F (180°C) for 25 minutes. Remove from oven and cool for 5 minutes, inverting cakes onto a wire rack. When completely cool, spread 1 cup (250 mL) of jam on top of first layer and sandwich with second layer. Spread remaining jam on top of cake and frost sides with whipped cream.

Variation: For a special garnish, dip dried apricots halfway into melted chocolate. Press chocolate apricots into whipped cream, around sides of cake.

CHOCOLATE BANANA CAKE

They'll go bananas over this one.

⅔ cup	milk	150 mL
1 tbsp.	vinegar	15 mL
2 cups	flour	500 mL
¾ cup	cocoa	175 mL
1½ tsp.	baking powder	7 mL
1 tsp.	baking soda	5 mL
¾ tsp.	salt	3 mL
¾ cup	butter OR margarine	175 mL
1⅔ cups	sugar	400 mL
2	eggs	2
1¼ cups	mashed banana	300 mL
1 tsp.	vanilla	5 mL
1	banana, sliced (optional)	1

Combine milk and vinegar; set aside. In a separate bowl, combine flour, cocoa, baking powder, baking soda and salt. In another bowl, cream butter and sugar. Beat in eggs, 1 at a time, and then stir in banana and vanilla. Add dry ingredients, alternately with soured milk, to the creamed mixture, mixing lightly after each addition. Spoon batter into 2 greased 9″ (23 cm) round layer pans. Bake at 350°F (180°C) for 40 to 45 minutes. Remove from oven and cool for 10 minutes, inverting cakes onto a wire rack. When completely cool, prepare a double recipe of Chocolate Buttercream Frosting, page 11. Spread a ½″ (1.3 cm) layer of frosting on first cake layer and sandwich with second layer. Spread remaining frosting on top and sides of cake. Just before serving, garnish top of cake with banana slices.

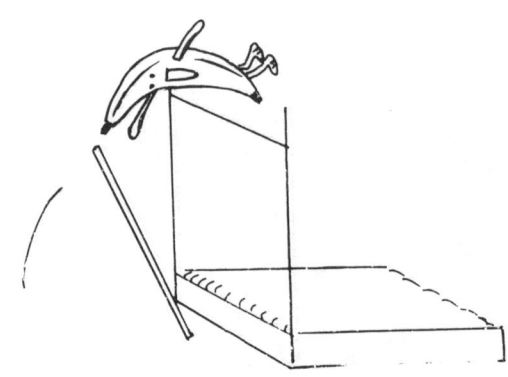

CHERRY CHOCOLATE UPSIDE DOWN CAKE

You'll flip over this dessert!

19 oz.	can cherry pie filling	540 mL
1¾ cups	flour	425 mL
1½ cups	sugar	375 mL
½ cup	cocoa	125 mL
2 tsp.	baking powder	10 mL
½ tsp.	salt	2 mL
1 cup	milk	250 mL
¾ cup	butter OR margarine	175 mL
4	eggs	4
1 tsp.	vanilla	5 mL
	whipped cream (optional)	

Spread pie filling evenly over bottom of a greased 9″ x 13″ (23 cm x 33 cm) pan. In a bowl, combine flour, sugar, cocoa, baking powder and salt. Add milk, butter, eggs and vanilla. Beat well for 2 minutes. Pour batter over pie filling. Bake at 350°F (180°C) for 40 to 45 minutes. Remove from oven and let stand for 10 minutes. Invert and cool. Serve with whipped cream, if desired.

Variation: Try peach pie filling for a luscious change.

 In 1657 the first Chocolate Houses appeared in England to rival the fashionable coffee shops.

CHOCOLATE-FLECKED APPLE CAKE

A delect-"apple" dessert!

1¾ cups	flour	425 mL
1 cup	sugar	250 mL
½ tsp.	salt	2 mL
½ tsp.	baking soda	2 mL
2	eggs	2
½ cup	oil	125 mL
½ cup	apple juice	125 mL
1 tsp.	vanilla	5 mL
1½ cups	peeled and chopped apple	375 mL
¾ cup	chopped chocolate chips	175 mL
½ cup	chopped walnuts	125 mL

Combine flour, sugar, salt and baking soda. In a separate bowl, combine eggs, oil, apple juice and vanilla. Add dry ingredients to apple juice mixture and stir until mixed. Add apples, chocolate and walnuts and mix again. Pour into greased 9" (23 cm) square pan and bake at 350°F (180°C) for 40 to 45 minutes. Frost with Chocolate Buttercream Frosting, page 11.

 Spanish princess Maria-Theresa of Spain married King Louis XIV in 1659. She introduced hot chocolate to the French Court. The ladies of the Court loved the drink but when they heard of its love powers the demand for it skyrocketed.

ORANGE CHOCOLATE CAKE

What a wonderful flavor combination!

1 cup	sugar	250 mL
1 cup	mayonnaise	250 mL
1 tbsp.	grated orange rind	15 mL
1 cup	orange juice	250 mL
1 tsp.	vanilla	5 mL
2 cups	flour	500 mL
½ tsp.	salt	2 mL
2 tsp.	baking soda	10 mL
¼ cup	cocoa	50 mL

Cream together sugar and mayonnaise. Add orange rind, orange juice and vanilla and mix thoroughly. In a separate bowl, combine flour, salt, baking soda and cocoa. Stir dry ingredients into the creamed mixture. Pour into greased 9″ x 13″ (23 cm x 33 cm) pan and bake at 350°F (180°C) for 40 to 45 minutes. Cool and frost with Chocolate Orange Frosting, recipe follows.

Note: You may also garnish this cake with thin orange slices, split halfway and twisted.

CHOCOLATE ORANGE FROSTING

1½ cups	icing (confectioners') sugar	375 mL
¼ cup	butter OR margarine	50 mL
¼ cup	cocoa	50 mL
1 tbsp.	grated orange rind	15 mL
3 tbsp.	orange juice	45 mL

Combine all ingredients and beat with electric mixer until light and fluffy.

ORANGE LIQUEUR MICROWAVE CAKE

This cake received rave reviews. I didn't tell them how easy it was to prepare.

½ cup	chocolate chips	125 mL
¾ cup	flour	175 mL
½ tsp.	baking powder	2 mL
½ tsp.	baking soda	2 mL
¼ tsp.	salt	1 mL
½ cup	butter OR margarine	125 mL
½ cup	sugar	125 mL
2	eggs	2
½ tsp.	vanilla	2 mL
1 tbsp.	grated orange rind	15 mL
1 tbsp.	orange liqueur	15 mL
½ cup	sour cream	125 mL

Melt chocolate chips in a small bowl on MEDIUM (50%) power, about 3 minutes. Stir to melt. In a separate bowl, combine flour, baking powder, baking soda and salt. In another bowl, cream butter and sugar. Add eggs, vanilla, orange rind and liqueur. Beat until light. Stir in melted chocolate. Add dry ingredients, alternately with sour cream, to the creamed mixture. Stir until smooth. Pour into an 8" (20 cm) square glass pan lined with wax paper. Bake at MEDIUM (50%) power for 4 minutes. Rotate dish ¼ turn every minute. Bake at HIGH (100%) power for 3 minutes. Rotate dish 3 times. Let stand 10 to 15 minutes. Invert pan, remove wax paper and cool. Frost with Orange Liqueur Frosting, recipe follows.

▲ ORANGE LIQUEUR FROSTING ▲

½ cup	chocolate chips	125 mL
2 tbsp.	butter OR margarine	30 mL
½ cup	icing (confectioners') sugar	125 mL
1 tbsp.	orange liqueur	15 mL
1 - 2 tbsp.	milk	15 - 30 mL

Melt chocolate chips on MEDIUM (50%) power, for 3 minutes. Beat in butter, icing sugar, liqueur and enough milk to make frosting spreadable.

PINEAPPLE CHOCOLATE CAKE

A very moist, uniquely flavored cake.

¾ cup	flour	175 mL
⅔ cup	sugar	150 mL
½ tsp.	baking soda	2 mL
½ tsp.	salt	2 mL
1 tsp.	vanilla	5 mL
1 oz.	unsweetened chocolate, melted (1 square)	30 g
⅓ cup	shortening	75 mL
¼ cup	milk	50 mL
3 tbsp.	pineapple juice	45 mL
½ cup	crushed, drained pineapple	125 mL
2	eggs	2

Place all ingredients, except 1 egg, in a large bowl. Blend with electric mixer on low, then beat for 1 minute on medium. Add remaining egg. Beat 1 more minute on medium. Spread batter in a 9'' (23 cm) round pan lined with wax paper. Microwave at MEDIUM (50%) power for 6 minutes, rotating ¼ turn every 2 minutes. Increase power to HIGH (100%), and microwave 2 - 3 more minutes. Let stand 5 -10 minutes. Invert pan, remove wax paper and cool completely. Frost with Pineapple Chocolate Frosting, recipe follows.

PINEAPPLE CHOCOLATE FROSTING

1 oz.	unsweetened chocolate, melted (1 square)	30 g
2 tbsp.	butter OR margarine	30 mL
3 tbsp.	pineapple juice	45 mL
½ tsp.	vanilla	2 mL
2 cups	icing (confectioners') sugar	500 mL
⅛ tsp.	salt	0.5 mL
	pineapple tidbits for garnish (optional)	

Melt chocolate on MEDIUM (50%) power. Combine all ingredients and beat with an electric mixer until light and fluffy. Spread over cake and garnish with pineapple tidbits, if desired.

CHOCOLATE PEAR FLAN

This looks spectacular and tastes even better.

1½ cups	graham wafer crumbs	375 mL
⅓ cup	butter OR margarine, melted	75 mL
¾ cup	finely chopped, toasted hazelnuts	175 mL
8 oz.	cream cheese	250 g
¼ cup	sugar	50 mL
1	egg	1
3 oz.	semisweet chocolate, melted (3 squares)	85 g
3 - 4	fresh pears, peeled, cored and halved OR 14 oz. (398 mL) can pear halves	3 - 4
1 oz.	semisweet chocolate (1 square)	30 g
1 tbsp.	butter OR margarine	15 mL
¼ cup	toasted hazelnuts, coarsely chopped whipped cream for garnish (optional)	50 mL

Combine graham crumbs, melted butter and finely chopped hazelnuts. Press firmly into 9" (23 cm) flan pan. Beat cream cheese and sugar until smoothly blended. Add egg and 3 oz. (85 g) melted chocolate. Mix thoroughly. Spread mixture evenly over crust. Drain pears and pat dry. Arrange pear halves over cream cheese layer. Bake at 375°F (190°C) for 25 minutes. Melt remaining chocolate with butter and drizzle over cooled flan. Sprinkle with coarsely chopped hazelnuts. Chill for 3 hours. Garnish with whipped cream, if desired, before serving.

 CHOCOLATE FONDUE

Great for entertaining, also a special treat for children.

6 oz.	**milk chocolate ***	**170 g**
6 oz.	**semisweet chocolate ***	**170 g**
¾ cup	**vanilla ice cream**	**175 mL**
2 tbsp.	**liqueur, any flavor (optional)**	**30 mL**
	bite-sized portions of apple, banana,	
	grapes, maraschino cherries,	
	orange, pear, pineapple,	
	strawberries OR any of your	
	favorite fruits	

In a fondue pot OR double boiler, break chocolate into small pieces. Add ice cream and liqueur and melt slowly, stirring constantly. Cool sauce slightly before serving as chocolate tends to run off fruit if too hot. With a long-handled fork, dip your favorite fruits into the sauce. Makes about 1½ cups (375 mL).

* Belgian, Dutch and Swiss chocolate are excellent choices. They may be purchased at chocolate specialty shops.

In 1660 a book was published in Paris recommending chocolate to "those who had the misfortune of finding themselves afflicted with the most universal of maladies occasioned by love."

 PINEAPPLE CHOCOLATE FUDGE

A tasty twist to an old favorite!

2 cups	sugar	500 mL
1 cup	brown sugar	250 mL
1 cup	crushed pineapple with juice	250 mL
½ cup	milk	125 mL
1 tbsp.	corn syrup	15 mL
¼ tsp.	salt	1 mL
2 tbsp.	butter OR margarine	30 mL
3 cups	miniature marshmallows	750 mL
1 tsp.	vanilla	5 mL
1 cup	chopped nuts	250 mL
1 cup	chocolate chips	250 mL

In a saucepan, combine sugars, pineapple with juice, milk, syrup, salt and butter. Cook, over medium heat, until the soft-ball stage OR 236°F (120°C), stirring occasionally. Remove from heat and add marshmallows and vanilla. Mix until marshmallows are melted, then beat until mixture becomes heavy and creamy. Add nuts and chocolate chips. Spread into a greased 8″ (20 cm) square pan.

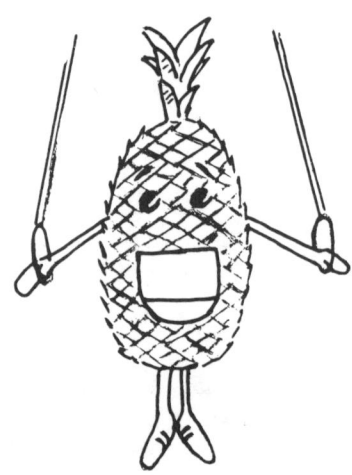

 English Quakers praised chocolate as a healthful substitute for gin.

♦ CHOCOLATE ORANGE MOUSSE ♦

Simply scrumptious!

3 cups	miniature marshmallows	750 mL
½ cup	milk	125 mL
1 cup	chocolate chips	250 mL
2 tbsp.	orange liqueur	30 mL
1 tsp.	vanilla	5 mL
1½ cups	whipping cream	375 mL
6	orange segments	6
	fresh mint leaves (optional)	

In a saucepan, combine marshmallows, milk and chocolate chips. Stir, over low heat, until completely melted. Remove from heat and stir in liqueur and vanilla. Set aside to cool. Whip cream until stiff. Fold into chocolate mixture. Spoon into 6 parfait glasses and chill. Garnish with orange segments and mint leaves, if desired. Serves 6.

♦ *In 1685, Queen Maria-Theresa's doctor, Dr. Bachot, said, "Well-made chocolate is such a noble invention that it, rather than nectar and ambrosia, should be known as the food of the gods."*

STEAMED CHOCOLATE RAISIN PUDDING

A mouth-watering treat!

⅓ cup	butter OR margarine	75 mL
⅓ cup	cocoa	75 mL
½ cup	boiling water	125 mL
1 cup	sugar	250 mL
1	egg, beaten	1
1 cup	flour	250 mL
½ tsp.	salt	2 mL
1 tsp.	baking powder	5 mL
¼ cup	milk	50 mL
1 tsp.	vanilla	5 mL
1 cup	raisins	250 mL

Place butter and cocoa in a large bowl. Pour boiling water over and combine. Add sugar and egg, stirring until smooth. Sift together flour, salt and baking powder. Add flour mixture, alternately with milk, to the cocoa mixture. Stir in vanilla and raisins. Pour into greased 1.5-quart (1.5 L) casserole. Cover and bake at 400°F (200°C) for 45 minutes. Serve warm with vanilla ice cream.

Note: For a special treat, raisins may be plumped in ¼ cup (50 mL) of rum or brandy for half an hour and/or pudding may be topped with a Rum-Raisin Sauce, recipe follows.

RUM-RAISIN SAUCE

½ cup	rum	125 mL
½ cup	seedless raisins	125 mL
½ cup	sugar	125 mL
¼ cup	water	50 mL
½ tsp.	cinnamon	2 mL
2 tbsp.	grated lemon rind	30 mL
¼ cup	chopped nuts	50 mL

Soak raisins in rum until plump. In a saucepan, combine sugar, water and cinnamon. Bring to a boil, over medium heat, and boil for 2 minutes. Add raisins and rum and cook 5 minutes more. Remove from heat, add lemon rind and nuts, and mix thoroughly. Serve warm. Makes 1¼ cups (300 mL).

VEGETABLES

CHOCOLATE YAM LOAF

This loaf will be gone in no time!

2	eggs	2
1½ cups	sugar	375 mL
½ cup	oil	125 mL
1 cup	cooked and mashed yams *	250 mL
1½ cups	flour	375 mL
1 tsp.	baking soda	5 mL
½ tsp.	salt	2 mL
½ tsp.	baking powder	2 mL
1 tsp.	cinnamon	5 mL
⅓ cup	orange juice	75 mL
½ cup	finely chopped walnuts	125 mL
⅓ cup	grated semisweet chocolate	75 mL

Combine eggs, sugar and oil. Beat until light and fluffy. Blend in yams. In a separate bowl, combine flour, baking soda, salt, baking powder and cinnamon. Add dry ingredients, alternately with orange juice, to yam mixture. Blend until smooth. Fold walnuts and grated chocolate into batter. Spread into a greased 5" x 9" (13 cm x 23 cm) loaf pan and bake at 350°F (180°C) for 45 to 50 minutes. Makes 1 loaf. This loaf may be served as is or frosted with Chocolate Orange Frosting, page 47.

*Yams are often confused with sweet potatoes, especially the brilliant orange variety of sweet potato which is softer, sweeter and more moist.

You may substitute cooked, mashed sweet potato or pumpkin for the yams.

▲ CHOCOLATE ZUCCHINI LOAF ▲

This is a heavy loaf but it is very moist and very good.

1 cup	**butter OR margarine**	250 mL
3 cups	**sugar**	750 mL
4	**eggs**	4
1 cup	**oil**	250 mL
1 cup	**sour milk ***	250 mL
2 tsp.	**vanilla**	10 mL
4 cups	**grated zucchini**	1 L
5 cups	**flour**	1.25 L
½ cup	**cocoa plus 2 tsp.**	135 mL
2 tsp.	**salt**	10 mL
2 tsp.	**baking soda**	10 mL
1½ cups	**chocolate chips**	375 mL

Cream butter and sugar then beat in eggs and oil. Stir in milk, vanilla and zucchini. Add flour, cocoa, salt and baking soda and mix well. Divide batter into 3, 5″ x 9″ (13 cm x 23 cm) loaf pans and sprinkle ½ cup (125 mL) of chocolate chips on top of each loaf. Bake at 325°F (160°C) for 75 minutes. Makes 3 loaves.

* To sour milk, put 1 tbsp. (15 mL) vinegar or lemon juice in a measuring cup. Fill cup with milk until 1 cup (250 mL) level is reached.

▲ *In 1705 Dr. Stephani Blancardi of Amsterdam said, "Tasty chocolate is also a veritable balm of the mouth, for the maintaining of all glands and humors in a good state of health. Thus it is, that all who do drink it possess a sweet breath."*

CHOCOLATE CARROT COOKIES

These soft, chewy cookies are especially good on the second day!

⅔ cup	butter OR margarine	150 mL
1 cup	brown sugar	250 mL
2	eggs	2
1 tsp.	vanilla	5 mL
4 tbsp.	milk	60 mL
½ cup	grated carrots	125 mL
2 cups	flour	500 mL
½ cup	cocoa	125 mL
½ tsp.	baking soda	2 mL
½ tsp.	salt	2 mL
½ cup	chopped walnuts	125 mL

Cream together butter and brown sugar. Add eggs, vanilla, milk and carrots. Mix thoroughly. Stir in flour, cocoa, baking soda and salt. Add walnuts and blend well. Drop by spoonfuls onto a greased baking sheet and bake at 350°F (180°C) for 15 minutes. Makes 30 cookies.

See photograph on back cover.

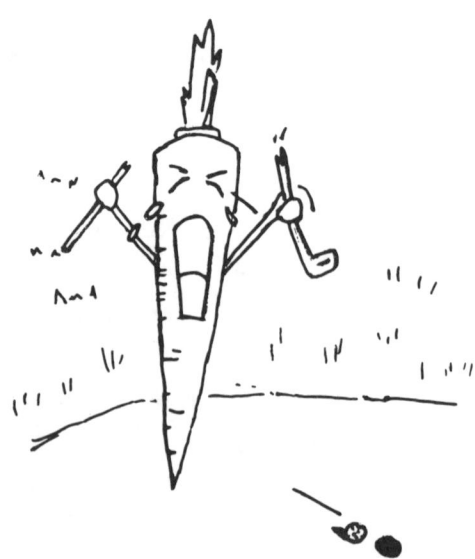

 The tradition of giving chocolate as a Valentine's Day gift was started in the mid-1700's by the famous lover, Casanova.

CHOCOLATE PUMPKIN COOKIES

A very moist batter.... A very moist cookie!

½ cup	butter OR margarine	125 mL
1 cup	sugar	250 mL
2	eggs	2
1 cup	cooked and mashed pumpkin	250 mL
1 tsp.	vanilla	5 mL
2 cups	flour	500 mL
⅓ cup	cocoa	75 mL
1½ tsp.	baking soda	7 mL
½ tsp.	salt	2 mL
½ tsp.	cinnamon	2 mL

Cream together butter and sugar. Add eggs, pumpkin and vanilla and blend well. Stir in flour, cocoa, baking soda, salt and cinnamon. Drop batter by spoonfuls onto greased baking sheet. Bake at 350°F (180°C) for 15 minutes. Makes 30 cookies.

 In 1765 the first chocolate factory appeared in the U.S.A.

PUMPKIN CHIP COOKIES

Pumpkin, orange and chocolate chips - great!

1 cup	sugar	250 mL
½ cup	butter OR margarine	125 mL
1 cup	cooked OR canned pumpkin	250 mL
1 tbsp.	grated orange rind	15 mL
½ tsp.	vanilla	2 mL
2 cups	flour	500 mL
1 tsp.	baking powder	5 mL
1 tsp.	baking soda	5 mL
1 tsp.	cinnamon	5 mL
¼ tsp.	salt	1 mL
1 cup	chocolate chips	250 mL

Cream together sugar and butter. Add pumpkin, orange rind and vanilla and blend well. Stir in flour, baking powder, baking soda, cinnamon and salt. Add chocolate chips and mix thoroughly. Drop batter by spoonfuls onto greased baking sheet and bake at 375°F (190°C) for 12 to 15 minutes. Makes 30 cookies.

 In 1775 Swedish naturalist Carolus Linnaeus gave the cocoa tree its botanical name, "theobroma," Greek for "food of the gods."

 CORN CRACKLES

Children love these — adults too!

4 cups	popped corn	1 L
2 cups	crushed cornflakes	500 mL
1 cup	chopped peanuts	250 mL
2 cups	chocolate chips	500 mL
⅓ cup	butter OR margarine	75 mL

In a large bowl, combine popcorn, cornflakes and peanuts. In a saucepan, over low heat, melt together chocolate chips and butter. When completely melted, remove from heat and pour over corn mixture. Stir until evenly coated. Spread mixture into greased 9″ x 13″ (23 cm x 33 cm) pan. Chill and cut into squares.

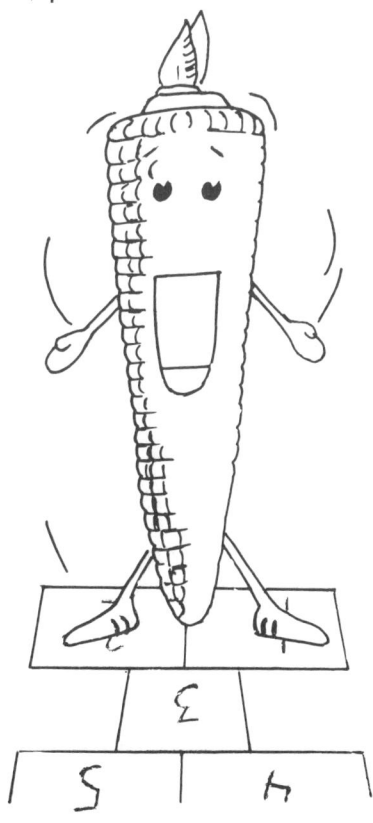

▲ *Thomas Jefferson said, "The superiority of chocolate, both for health and nourishment, will soon give it preference over tea and coffee in America as it has in Spain."*

◣ CHOCOLATE CARROT CAKE ◣

Chocolate! and good for you too!

1 cup	sugar	250 mL
3	eggs	3
1 cup	oil	250 mL
⅓ cup	cocoa	75 mL
1⅓ cups	flour	325 mL
½ tsp.	salt	2 mL
1 tsp.	baking soda	5 mL
1 tsp.	cinnamon	5 mL
2 cups	grated carrots	500 mL
½ cup	chopped walnuts	125 mL

Cream together sugar and eggs; add oil and mix until blended. In a separate bowl, combine cocoa, flour, salt, baking soda and cinnamon. Stir dry ingredients into creamed mixture. Add carrots and walnuts and blend thoroughly. Pour into ungreased 8″ (20 cm) square pan and bake at 300°F (150°C) for 55 to 60 minutes. Frost with Cream Cheese Frosting, recipe follows.

Variation: Bake batter in a 5″ x 9″ (13 cm x 23 cm) loaf pan for 75 to 80 minutes. Split cooled loaf into 3 layers. Frost only between layers and on top of cake with Cream Cheese Frosting. Garnish top of cake with orange slice twists.

◣ CREAM CHEESE FROSTING ◣

3 oz.	cream cheese, softened	85 g
2 tbsp.	butter OR margarine, softened	30 mL
1½ cups	icing (confectioners') sugar	375 mL
½ tsp.	vanilla	2 mL
1-2 tbsp.	milk	15-30 mL
½ cup	chopped walnuts (optional)	125 mL

Beat together cream cheese and butter until light and fluffy. Gradually sift in icing sugar. Stir in vanilla and add enough milk to make frosting spreadable. Garnish with chopped walnuts, if desired.

CHOCOLATE POTATO CAKE

Moist and delicious!

⅔ cup	butter OR margarine	150 mL
1½ cups	sugar	375 mL
½ cup	whipping cream, warmed	125 mL
1 cup	hot mashed potatoes	250 mL
1 cup	melted chocolate chips	250 mL
3	eggs	3
1 tsp.	vanilla	5 mL
½ cup	milk	125 mL
2 tsp.	baking soda	10 mL
⅓ cup	warm water	75 mL
2 cups	flour	500 mL
2 tsp.	baking powder	10 mL
½ tsp.	salt	2 mL

Cream together butter and sugar. Pour warm cream into potatoes and add to the first mixture. Stir in melted chocolate, eggs, vanilla and milk. Dissolve baking soda in water and add to potato mixture. In a separate bowl, combine flour, baking powder and salt. Fold into potato mixture. Spoon batter into 2 greased 8″ (20 cm) layer pans and bake at 375°F (190°C) for 45 minutes. Remove from oven, let stand for 10 minutes and invert cakes onto a wire rack. While cooling, prepare Fudge Frosting, page 68. Spread a thick layer of frosting on first layer of cake and sandwich with second layer. Spread remaining frosting on top and sides of cake.

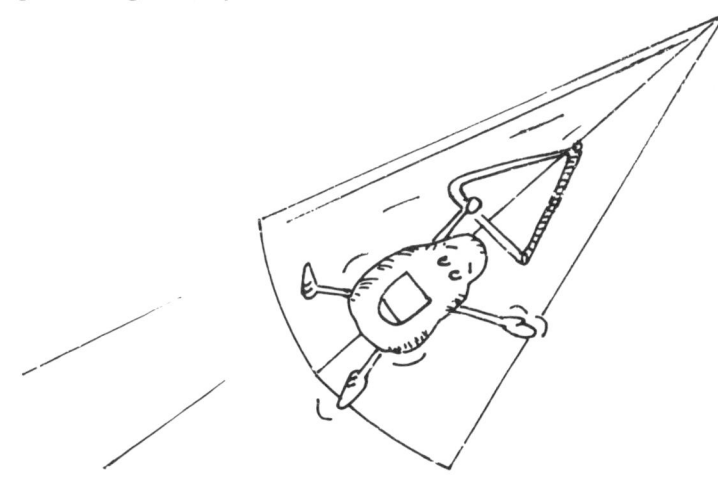

▲ POTATO FUDGE CAKE ▲

This cake is "spud"-tacular!

4	eggs, separated	4
1⅓ cups	butter OR margarine	325 mL
2 cups	sugar	500 mL
⅔ cup	cocoa	150 mL
½ cup	milk	125 mL
1 tbsp.	grated lemon rind	15 mL
2½ cups	flour	625 mL
1 tbsp.	baking powder	15 mL
¼ tsp.	salt	1 mL
½ tsp.	cinnamon	2 mL
1 cup	peeled, grated raw potato	250 mL
½ cup	chopped walnuts	125 mL

Beat egg whites until stiff. In a separate bowl, cream butter and sugar. Beat in egg yolks, cocoa, milk and lemon rind. In another bowl, combine flour, baking powder, salt and cinnamon. Stir dry ingredients into the creamed mixture. Mix in potatoes and walnuts and gently fold in egg whites. Spoon batter into greased bundt pan and bake at 325°F (160°C) for 75 minutes. Allow cake to cool for 20 minutes before inverting pan and removing cake. Serve with ice cream or, for a special occasion, drizzle with a thin Chocolate Glaze, recipe follows.

▲ CHOCOLATE GLAZE ▲

6 oz.	semisweet chocolate (6 squares)	170 g
⅓ cup	whipping cream, warmed	75 mL
2 tbsp.	rum OR apricot brandy OR coffee liqueur OR milk	30 mL

Melt chocolate over low heat or on MEDIUM (50%) power in your microwave for 3 to 4 minutes. Stir to finish melting chocolate. Stir in cream and rum. Drizzle warm glaze over cake.

CHOCOLATE RUTABAGA CAKE

Surprisingly good!

½ cup	butter OR margarine	125 mL
½ cup	oil	125 mL
2 cups	sugar	500 mL
2	eggs	2
1 tsp.	vanilla	5 mL
½ cup	sour milk, see page 57	125 mL
½ cup	cocoa	125 mL
½ tsp.	cinnamon	2 mL
½ tsp.	ground cloves	2 mL
½ tsp.	salt	2 mL
1 tsp.	baking soda	5 mL
2 cups	flour	500 mL
⅓ cup	grated raw rutabaga OR turnip *	75 mL
½ cup	chocolate chips	125 mL
½ cup	chopped walnuts	125 mL

Cream together butter, oil and sugar. Add eggs, vanilla and sour milk. Beat thoroughly. In a separate bowl, combine cocoa, cinnamon, cloves, salt, baking soda and flour. Add dry ingredients to creamed mixture and mix well. Fold in rutabaga. Spoon batter into greased 9″ x 13″ (23 cm x 33 cm) pan. Sprinkle chocolate chips and walnuts on top. Bake at 325°F (160°C) for 50 minutes.

* Rutabagas, also called yellow turnips, or Swedes, are very similar to turnips. Both may be mashed or baked. Turnips are great eaten raw, like apples. Rutabagas may be French-fried, like potatoes. They are usually larger, firmer and less watery than turnips.

CHOCOLATE SQUASH CAKE

Everyone asks for this recipe!

1¾ cups	butter OR margarine	425 mL
2 cups	sugar	500 mL
3	eggs	3
2 tsp.	vanilla	10 mL
	rind and juice of ¼ lemon	
½ cup	milk	125 mL
2 cups	coarsely shredded squash	500 mL
2½ cups	flour	625 mL
9 tbsp.	cocoa	135 mL
2½ tsp.	baking powder	12 mL
1½ tsp.	baking soda	7 mL
¼ tsp.	salt	1 mL

Cream together butter and sugar. Beat in eggs, vanilla, lemon and milk then stir in squash. In a separate bowl, combine flour, cocoa, baking powder, baking soda and salt. Fold into the squash mixture until the dry ingredients are thoroughly moistened. Pour batter into greased tube or bundt pan. Bake at 350°F (180°C) for 1 hour. Allow cake to cool for 20 minutes. Invert pan and remove cake from pan. Top with Milk Glaze, page 30, or, for a double-chocolate treat, frost with Chocolate Buttercream Frosting, page 11.

 In 1828 Dutch chemist Conrad J. van Houten invented the cocoa press, improving the quality of chocolate by squeezing out part of the cocoa butter. This gave drinking chocolate a smoother consistency and a more pleasant flavor.

CHOCOLATE ZUCCHINI CAKE

Dark and delicious!

½ cup	butter OR margarine	125 mL
1¾ cups	sugar	425 mL
2	eggs	2
½ cup	oil	125 mL
1 tsp.	vanilla	5 mL
½ cup	sour milk, see page 57	125 mL
1 tsp.	baking soda	5 mL
6 tbsp.	cocoa	90 mL
½ tsp.	salt	2 mL
2½ cups	flour	625 mL
2 cups	shredded zucchini	500 mL
½ cup	chocolate chips	125 mL
½ cup	nuts	125 mL

Cream butter and sugar. Beat in eggs, oil, vanilla and sour milk. In a separate bowl, combine baking soda, cocoa, salt and flour and add to the creamed mixture. Stir in zucchini. Spoon batter into greased 9″ x 13″ (23 x 33 cm) pan. Sprinkle with chocolate chips and nuts. Bake at 325°F (160°C) for 40 to 50 minutes.

In 1847 the British firm of Fry and Sons mixed cocoa butter with chocolate liquor and sugar to form the first solid "eating" chocolate.

SAUERKRAUT CHOCOLATE CAKE

Another personal favorite!

2¼ cups	flour	550 mL
½ cup	cocoa	125 mL
1 tsp.	baking powder	5 mL
1 tsp.	baking soda	5 mL
¼ tsp.	salt	1 mL
⅔ cup	butter OR margarine	150 mL
1½ cups	sugar	375 mL
3	eggs	3
1 tsp.	vanilla	5 mL
1 cup	strong coffee	250 mL
⅔ cup	rinsed, drained and coarsely chopped sauerkraut	150 mL

Combine flour, cocoa, baking powder, baking soda and salt. In a separate bowl, cream butter and gradually add sugar, eggs, vanilla and coffee; mix well. Add creamed mixture and dry ingredients alternately into a third bowl. Stir in sauerkraut. Pour into 2 greased 8″ (20 cm) round pans. Bake at 350°F (180°C) for 25 to 30 minutes. Frost with Fudge Frosting, recipe follows.

See photograph, opposite.

▲ FUDGE FROSTING ▲

2 cups	sugar	500 mL
1 cup	water	250 mL
2 oz.	unsweetened chocolate (2 squares)	55 g
2 tbsp.	corn syrup	30 mL
¼ tsp.	salt	1 mL
2 tbsp.	butter OR margarine	30 mL
1 tsp.	vanilla	5 mL

In a saucepan, combine sugar, water, chocolate, syrup and salt. Stir, over low heat, until sugar dissolves. Cook to softball stage, 236°F (120°C). Remove from heat and add butter. Cool to lukewarm, then add vanilla and beat to a spreading consistency. If mixture is too stiff, reheat or add 1 tsp. (5 mL) of oil and reheat.

Sauerkraut Chocolate Cake, page 68.
Fudge Frosting, page 68.

CHOCO-MINT BEET CAKE

This cake is a real taste experience!

1½ cups	sugar	375 mL
2	eggs	2
1 cup	oil	250 mL
19 oz.	can beets, drained and puréed	540 mL
1 tsp.	vanilla	5 mL
¾ tsp.	pure peppermint extract	3 mL
½ cup	cocoa	125 mL
1¾ cups	flour	425 mL
1½ tsp.	baking soda	7 mL
¼ tsp.	salt	1 mL

In a large bowl, combine sugar and eggs. Add oil, beets, vanilla and peppermint extract. Mix well. Stir in cocoa, flour, baking soda and salt. Pour batter into greased 9'' (23 cm) square pan. Bake at 350°F (180°C) for 45 to 50 minutes. Cool on cake rack and frost with Chocolate Buttercream Frosting, page 11, in which ¼ tsp. (1 mL) of pure peppermint extract has been substituted for vanilla.

 In 1876 candy maker Daniel Peter and chemist Henri Nestlé pooled their efforts and invented milk chocolate for eating. It was dry and lumpy but it was SOLID.

PARSNIP SPICE CAKE

Rich and spicy!

½ cup	butter OR margarine	125 mL
1½ cups	sugar	375 mL
2	eggs	2
⅓ cup	grated parsnips	75 mL
½ cup	water	125 mL
1 cup	sour milk, see page 57	250 mL
1 tsp.	vanilla	5 mL
2 cups	flour	500 mL
1½ tsp.	baking powder	7 mL
1 tsp.	baking soda	5 mL
½ tsp.	salt	2 mL
½ cup	cocoa	125 mL
2 tsp.	instant coffee granules	10 mL
1 tsp.	cinnamon	5 mL
½ tsp.	ginger	2 mL

Cream together butter and sugar. Beat in eggs then parsnips. Add water, sour milk, and vanilla and blend well. In a separate bowl, combine flour, baking powder, baking soda, salt, cocoa, coffee, cinnamon and ginger. Add dry ingredients to creamed mixture and mix thoroughly. Pour batter into greased 9" (23 cm) square pan and bake at 350°F (180°C) for 55 to 60 minutes. Frost with Cream Cheese Frosting, page 62.

CHOCOLATE POTATO BALLS

These are a little fussy to make, but worth it!

½ cup	cold mashed potatoes	125 mL
1½ cups	icing (confectioners') sugar	375 mL
2½ cups	finely shredded coconut	625 mL
½ tsp.	almond flavoring	2 mL
1 cup	chocolate chips	250 mL
2 tbsp.	butter OR margarine	30 mL

Combine potatoes, icing sugar, coconut and almond flavoring. Mix well. Shape potato mixture into small balls. Melt together chocolate chips and butter. With a fork, dip potato balls into chocolate and set on waxed paper to dry. If desired , dust with powdered sugar or candy sprinkles before chocolate has completely set. Makes about 18 potato balls.

 The Baby Ruth chocolate bar was not named for the famous baseball player, Babe Ruth; rather, it was named for the youngest daughter of U.S. President Grover Cleveland.

CHOCOLATE SPUDNUTS

Let these sit for a day or two, for flavors to ripen.

2½ cups	flour	625 mL
2 tsp.	baking powder	10 mL
1 tsp.	baking soda	5 mL
½ tsp.	salt	2 mL
½ cup	cocoa	125 mL
¾ cup	chocolate chips, melted	175 mL
½ cup	warm mashed potatoes	125 mL
1 cup	sugar	250 mL
½ cup	milk	125 mL
1 tsp.	vanilla	5 mL
1	egg, beaten	1

Combine flour, baking powder, baking soda, salt and cocoa. In a separate bowl, combine melted chocolate and potatoes. Stir in sugar, milk, vanilla and egg. Mix until smooth. Pour liquid ingredients into dry, stirring only to moisten flour. Place dough on floured board and knead 2 or 3 times. Roll out dough to ¼" (1 cm) thickness. Cut with floured doughnut cutter. Let stand 10 minutes. Deep-fry in oil at 375°F (190°C) for 1 minute per side. Remove from oil and drain on paper towels. Dip in icing sugar OR top with Chocolate Glaze, page 64, if desired. Makes 30 doughnuts.

BREADS &
CEREALS

▲ CHOCOLATE BREAD ▲

Serve this toasted with butter, cream cheese OR marmalade.

1 cup	milk	250 mL
1¼ cups	water	300 mL
1 tbsp.	yeast (7 g envelope)	15 mL
2 tbsp.	oil	30 mL
⅔ cup	sugar	150 mL
½ tsp.	vanilla	2 mL
6 cups	flour	1.5 L
¾ cup	cocoa	175 mL
2 tsp.	salt	10 mL
1 tsp.	cinnamon	5 mL

In a saucepan, combine milk and water. Heat until lukewarm then sprinkle yeast into mixture. Let sit for 5 minutes. Stir in oil, sugar and vanilla. In a medium-sized bowl, sift together flour, cocoa, salt and cinnamon. Stir half the dry mixture into the liquid ingredients. Gradually add remaining dry ingredients, stirring until the dough leaves the sides of the bowl. Turn out on a floured board and knead lightly until smooth (5 to 10 minutes). Place in a greased bowl and lightly grease the top of the dough. Cover with a towel. Let the dough rise until double in bulk (approximately 2 hours). Knead lightly again and let rise a second time. Divide dough and shape into 2 loaves. Place in 2, 5" x 9" (13 cm x 23 cm) greased loaf pans. Lightly grease the top of each loaf. Cover and let rise until double in bulk. Bake at 350°F (180°C) for 40 minutes. Remove from oven and brush top crusts with melted butter. Remove loaves from pans and cool. Makes 2 loaves.

▲ CHOCOLATE CINNAMON BUNS ▲

Serve warm with butter OR peanut butter.

1 tsp.	sugar	5 mL
1 cup	lukewarm water	250 mL
1 tbsp.	yeast (7 g envelope)	15 mL
¼ cup	shortening	50 mL
⅓ cup	sugar	75 mL
⅓ cup	cocoa	75 mL
1 tsp.	salt	5 mL
1	egg	1
2¼ cups	flour	550 mL
2 tbsp.	butter	30 mL
1½ tsp.	cinnamon	7 mL
¾ cup	brown sugar	175 mL

Dissolve sugar in water. Sprinkle in yeast and let stand for 10 minutes. Stir in shortening, sugar, cocoa, salt, egg and 1 cup (250 mL) of flour. Beat for 2 minutes with electric mixer on medium. Stir in remaining flour. Scrape sides of bowl and cover, allowing to rise for 1 hour. Turn out on a well-floured board and roll into a 9″ x 12″ (23 cm x 30 cm) rectangle. Spread butter, cinnamon and brown sugar onto dough. Roll dough up gently and pinch edge into roll to seal, then cut into 12 pieces. Place buns in greased 9″ (23 cm) square pan. Let rise in a warm place for 40 minutes. Bake at 375°F (190°C) for 25 minutes. Makes 12 buns.

CHOCOLATE BRAN MUFFINS

Delicious and healthy too!

1 cup	bran	250 mL
½ cup	oat bran	125 mL
1 cup	flour	250 mL
1½ tsp.	baking powder	7 mL
1 tsp.	baking soda	5 mL
½ tsp.	salt	2 mL
2 tsp.	grated orange rind	10 mL
2 oz.	semisweet chocolate, chopped (2 squares)	55 g
¼ cup	chopped pecans	50 mL
¼ cup	raisins	50 mL
⅓ cup	mashed banana	75 mL
¼ cup	oil	50 mL
¼ cup	molasses	50 mL
2 tbsp.	honey	30 mL
¼ cup	orange juice	50 mL
½ cup	yogurt	125 mL
2	eggs	2

Combine all dry ingredients. In a separate bowl, combine remaining ingredients. Pour liquid ingredients into the dry and stir just until moistened. Spoon batter into greased muffin tins and bake at 375°F (190°C) for 20 minutes. Makes 18 medium-sized muffins.

See photograph on back cover.

CHOCOLATE GRANOLA MUFFINS

A crunchy treat!

1½ cups	flour	375 mL
1 cup	brown sugar	250 mL
⅓ cup	cocoa	75 mL
1 tbsp.	baking powder	15 mL
1 tsp.	salt	5 mL
1 cup	granola cereal	250 mL
2	eggs	2
1 cup	milk	250 mL
⅓ cup	oil	75 mL
1 tsp.	vanilla	5 mL

Combine flour, sugar, cocoa, baking powder, salt and granola. In a separate bowl, beat together eggs, milk, oil and vanilla. Pour liquid ingredients into dry and stir only until moistened (batter should be lumpy). Spoon batter into greased muffin tins. Bake at 375°F (190°C) for 18 to 20 minutes. Makes 12 large muffins.

See photograph on back cover.

 Chocolate is easily digested and for centuries it has been highly valued as a food and quick-energy source by soldiers, athletes and travellers.

CHIP OATMEAL BROWNIES

Kids really go for this one!

½ cup	butter OR margarine	125 mL
¼ cup	white sugar	50 mL
¼ cup	brown sugar	50 mL
1	egg	1
1 tsp.	vanilla	5 mL
½ cup	flour	125 mL
½ tsp.	baking powder	2 mL
¼ tsp.	salt	1 mL
1 cup	oatmeal	250 mL
1½ cups	chocolate chips	375 mL

Cream together butter and sugars. Add egg and vanilla; mix well. Blend in flour, baking powder and salt. Stir in oatmeal and 1 cup (250 mL) of chocolate chips. Spread into greased 9″ (23 cm) square pan. Sprinkle remaining chocolate chips over the batter. Bake at 350°F (180°C) for 30 to 35 minutes.

See photograph on back cover.

 During World War II chocolate was considered extremely important in the nourishment of soldiers. The Government even allocated precious shipping space for the importation of cacao beans.

OATMEAL CHIP COOKIES

Our household is never without these!

½ cup	butter OR margarine	125 mL
½ cup	brown sugar	125 mL
½ cup	white sugar	125 mL
1	egg	1
1 tbsp.	warm water	15 mL
½ tsp.	baking soda	2 mL
½ tsp.	vanilla	2 mL
¾ cup	flour	175 mL
½ tsp.	salt	2 mL
1½ cups	rolled oats	375 mL
1 cup	chocolate chips	250 mL
½ cup	chopped nuts	125 mL

Cream together butter and sugars. Add egg, water, baking soda and vanilla. Beat well. Add flour, salt, rolled oats, chocolate chips and nuts. Blend until batter is thoroughly mixed. Drop batter by teaspoonfuls onto greased cookie sheet and bake at 325°F (160°C) for 15 minutes. Makes 24 cookies.

See photograph on back cover.

 In 1953 Sir Edmund Hillary and his teammates devoured pounds of chocolate while climbing up Mount Everest.

🔔 CHOCOLATE NOODLE COOKIES 🔔

Kids can make this one.

1 cup	chocolate chips	250 mL
1 cup	butterscotch chips	250 mL
4 oz.	can chow mein noodles	113 g
1 cup	coarsely chopped almonds	250 mL

In a saucepan, melt chocolate chips and butterscotch chips together over low heat. When completely melted, remove from heat and stir in noodles and almonds. Blend well. Drop mixture by spoonfuls onto wax-paper-lined cookie sheet. Refrigerate until firm. Makes 24 cookies.

🔔 CHOCOLATE SHORTBREAD 🔔

Shortbread never tasted so good!

2 cups	flour	500 mL
⅔ cup	cocoa	150 mL
1 cup	icing (confectioners') sugar	250 mL
1 cup	butter	250 mL
1 tsp.	vanilla	5 mL

Sift together flour, cocoa and icing sugar. Cut butter into bits and add to first mixture, cutting butter in as for pie dough. Add vanilla and mix well, working dough with hands, until dough sticks together. Roll dough out to ½'' (1.3 cm) thickness and cut into bars. Place on ungreased baking sheet and bake at 300°F (150°F) for 25 minutes. Makes 30, 2'' (5 cm) square pieces.

See photograph on back cover.

▲ BRAN FLAKE CRUNCHIES ▲

Ooey, gooey, good!

⅔ cup	brown sugar	150 mL
½ cup	corn syrup	125 mL
⅓ cup	crunchy peanut butter	75 mL
⅓ cup	melted butter OR margarine	75 mL
1 tsp.	vanilla	5 mL
4 cups	bran flakes	1 L
½ cup	chocolate chips	125 mL
⅓ cup	peanut butter	75 mL

Combine sugar, syrup, crunchy peanut butter, butter and vanilla. Stir in bran flakes. Press into greased 8″ (20 cm) square pan and bake at 375°F (190°C) for 5 minutes. Melt chocolate chips and peanut butter, over hot water. Spread chocolate mixture evenly over baked layer. Cool in the refrigerator then cut into bars.

▲ OATMEAL CHIP SQUARES ▲

These squares are always highly praised!

1 cup	melted butter OR margarine	250 mL
1 cup	brown sugar	250 mL
½ cup	maple syrup	125 mL
1 cup	granola cereal	250 mL
5 cups	oatmeal	1.25 L
1 cup	chocolate chips	250 mL
1 cup	butterscotch chips	250 mL
½ cup	peanut butter	125 mL
½ cup	chopped, blanched peanuts	125 mL

Combine butter, sugar and maple syrup. Add granola and oatmeal. Press into greased 9″ x 13″ (23 cm x 33 cm) pan. Bake at 350°F (180°C) for 15 to 20 minutes. Cool. Melt together chocolate chips, butterscotch chips and peanut butter, over hot water. Spread chocolate mixture on oatmeal base and sprinkle with peanuts. Chill and cut into bars.

CHOCOLATE OATMEAL SQUARES

These won't be around long enough to cool.

1½ cups	oatmeal	375 mL
1¼ cups	flour	300 mL
½ tsp.	salt	2 mL
½ tsp.	baking soda	2 mL
½ cup	butter OR margarine	125 mL
1 cup	brown sugar	250 mL
2	eggs	2
10 oz.	can sweetened, condensed milk	300 mL
½ cup	cocoa	125 mL
½ cup	chopped nuts	125 mL
2 tsp.	vanilla	10 mL

Stir together oatmeal, flour, salt and baking soda. In a separate bowl, cream butter and gradually stir in sugar and eggs. Pour dry ingredients into the creamed mixture and stir well. In another bowl, gradually stir milk into cocoa. Add nuts and vanilla. Stir well. Spread ⅔ of the oat mixure in the bottom of an ungreased 9″ (23 cm) square cake pan. Pour cocoa mixture on top. Drop small spoonfuls of remaining oat mixture over all. Bake at 350°F (180°C) for 40 to 45 minutes.

 The cacao tree will only grow a mere 20° north or south of the equator.

 CRISPY RICE SQUARES

Kids love to make and eat these.

4 cups	**crispy rice cereal**	**1 L**
½ cup	**shredded coconut**	**125 mL**
2 cups	**chocolate chips**	**500 mL**
½ cup	**chopped nuts**	**125 mL**
10 oz.	**can sweetened condensed milk**	**300 mL**

Crush cereal. Add coconut, chocolate chips and nuts. Add milk and stir until well combined. Pour mixture into greased 8″ (20 cm) pan and bake at 350°F (180°C) for 40 minutes. Cool and cut into squares.

The Ivory Coast in West Africa is the number one producer of cacao beans in the world.

 GRANOLA CHIP BARS

These bars are packed with good taste.

½ cup	butter OR margarine	125 mL
1 cup	brown sugar	250 mL
1	egg	1
1 tsp.	vanilla	5 mL
1⅓ cups	flour	325 mL
½ tsp.	baking soda	2 mL
½ tsp.	salt	2 mL
½ tsp.	cinnamon	2 mL
¼ cup	milk	50 mL
1⅔ cups	granola cereal	400 mL
1 cup	shredded coconut	250 mL
1 cup	chocolate chips	250 mL

Cream together butter, brown sugar, egg and vanilla. In a separate bowl, combine flour, baking soda, salt and cinnamon. Add dry ingredients to the creamed mixture. Stir in milk, granola, coconut and chocolate chips. Spread mixture evenly onto greased baking sheet. Bake at 350°F (180°C) for 20 to 25 minutes. Cool completely, invert pan and cut into bars.

 400 cacao beans are needed to make one pound of chocolate.

PEANUT BUTTER GRANOLA BARS

Nutrition has never tasted so good!

2 tbsp.	butter OR margarine	30 mL
½ cup	peanut butter	125 mL
1 cup	white sugar	250 mL
1 cup	brown sugar	250 mL
3	eggs, well beaten	3
1 tsp.	vanilla	5 mL
2 cups	flour	500 mL
1 tsp.	salt	5 mL
1 tsp.	baking soda	5 mL
1 cup	granola cereal	250 mL
1 cup	rolled oats	250 mL
½ cup	chocolate chips	125 mL

Cream together butter, peanut butter, and sugars. Add eggs and vanilla. In a separate bowl, combine remaining ingredients. Add dry ingredients to creamed mixture; mix thoroughly. Spread mixture evenly onto small greased baking sheet. Bake at 350°F (180°C) for 20 to 25 minutes. Cool and cut into bars.

See photograph on back cover.

 Chocolate itself is virtually free of salt and cholesterol.

A top-line treat.

1⅔ cups	graham cracker crumbs	400 mL
½ cup	flour	125 mL
½ tsp.	baking soda	2 mL
½ cup	melted butter OR margarine	125 mL
½ cup	brown sugar	125 mL
2	eggs	2
1 tbsp.	water	15 mL
1 tsp.	vanilla	5 mL
2 cups	chocolate chips	500 mL
1 cup	chopped nuts	250 mL

Thoroughly combine graham cracker crumbs, flour and baking soda. In a separate bowl, combine butter and brown sugar. Stir in eggs, water and vanilla. Gradually add graham cracker mixture and blend well. Stir in 1 cup (250 mL) chocolate chips and nuts. Spread in greased 9" x 13" (23 cm x 33 cm) pan and bake at 375°F (190°C) for 16 to 18 minutes. Remove from oven and immediately sprinkle remaining chocolate chips over warm surface. Let stand until chocolate softens and spreads evenly. Cool and cut into bars.

See photograph on back cover.

 Swiss people eat the most chocolate per capita, averaging over 21 pounds per person annually.

GERMAN CHOCOLATE OAT CAKE

A very moist, tasty cake!

1¼ cups	boiling water	300 mL
1 cup	oatmeal	250 mL
½ cup	butter OR margarine	125 mL
1 cup	chocolate chips	250 mL
1½ cups	flour	375 mL
1 cup	sugar	250 mL
1 tsp.	baking soda	5 mL
1 cup	brown sugar	250 mL
½ tsp.	salt	2 mL
3	eggs, beaten	3
1 tsp.	vanilla	5 mL

Pour boiling water over oatmeal; add butter and chocolate chips. Let stand 20 minutes. In a separate bowl, combine flour, sugar, baking soda, brown sugar and salt. Add eggs, vanilla and oatmeal mixture, beating thoroughly. Pour batter into greased 9" x 13" (23 cm x 33 cm) pan and bake at 350°F (180°C) for 35 minutes. While baking, prepare Broiled Frosting, recipe follows.

BROILED FROSTING

½ cup	butter OR margarine	125 mL
¼ cup	whipped cream	50 mL
¾ cup	brown sugar	175 mL
½ cup	chopped nuts	125 mL
½ cup	coconut	125 mL
½ tsp.	vanilla	2 mL

Combine all ingredients in a saucepan. Cook over low heat for 3 minutes or until thick. Pour over baked cake and broil until golden and bubbly. Watch carefully to make sure that frosting does not burn.

WHOLE-WHEAT CHOCOLATE CAKE

A cake with texture and taste.

1 cup	boiling water	250 mL
½ cup	oatmeal	125 mL
4 tbsp.	cocoa	60 mL
2	eggs	2
½ cup	oil	125 mL
1½ cups	brown sugar	375 mL
1 tsp.	vanilla	5 mL
½ cup	whole-wheat flour	125 mL
½ cup	all-purpose flour	125 mL
1 tsp.	baking powder	5 mL
½ tsp.	salt	2 mL
1 tsp.	baking soda	5 mL

Combine water, oatmeal and cocoa; set aside to cool. In a large bowl, beat eggs. Add oil, brown sugar and vanilla. Blend thoroughly. Stir in flours, baking powder, salt and baking soda. Add oatmeal mixture and blend well. Spoon batter into greased 9″ x 13″ (23 cm x 33 cm) pan and bake at 350°F (180°C) for 30 to 35 minutes. Frost with Chocolate Buttercream Frosting, page 11.

 Canada ranks an impressive 7th in the world for chocolate consumption on a per capita basis, averaging over 8 pounds per person annually.

WHOLE-WHEAT CHOCO-PEANUT CAKE

Chocolate and peanut butter! Who can resist?

¾ cup	butter OR margarine	175 mL
½ cup	chunky peanut butter	125 mL
1¾ cups	sugar	425 mL
2	eggs	2
1 tsp.	vanilla	5 mL
1 cup	water	250 mL
1 cup	whole-wheat flour	250 mL
1 cup	all-purpose flour	250 mL
½ cup	cocoa	125 mL
1 tsp.	baking soda	5 mL
¼ tsp.	salt	1 mL

Cream together butter, peanut butter and sugar. Beat in eggs, vanilla and water. In a separate bowl, combine flours, cocoa, baking soda and salt. Beat dry ingredients into the creamed mixture. Pour batter into greased bundt pan and bake at 350°F (180°C) for 60 minutes. Cool cake for 20 minutes before inverting pan to turn out finished cake. Top with Chocolate Glaze, page 64, or frost with Chocolate Buttercream Frosting, page 11, in which ⅓ cup (75 mL) of peanut butter is substituted for butter.

 Due to the complex manufacturing process, it takes from two to four days just to make an individual-sized chocolate bar.

CHOCOLATE RICE PUDDING

A new taste treat!

1 cup	short-grain rice	250 mL
2 cups	water	500 mL
⅔ cup	sugar	150 mL
3 tbsp.	cocoa	45 mL
1	egg, beaten	1
1 tsp.	vanilla	5 mL
½ tsp.	salt	2 mL
¾ cup	milk	175 mL

In a saucepan, combine rice and water. Bring to a boil, over high heat, then simmer until rice is tender, about 15 minutes. In a bowl, combine sugar and cocoa. Add egg, vanilla and salt, then stir in milk. Pour cocoa mixture into cooked rice and increase heat to high. Stir constantly until mixture thickens and begins to boil, watching carefully so that mixture does not burn. Let boil for 1 minute then remove from heat. Cover for 20 minutes. Serve warm, topped with vanilla ice cream. Makes 6 servings.

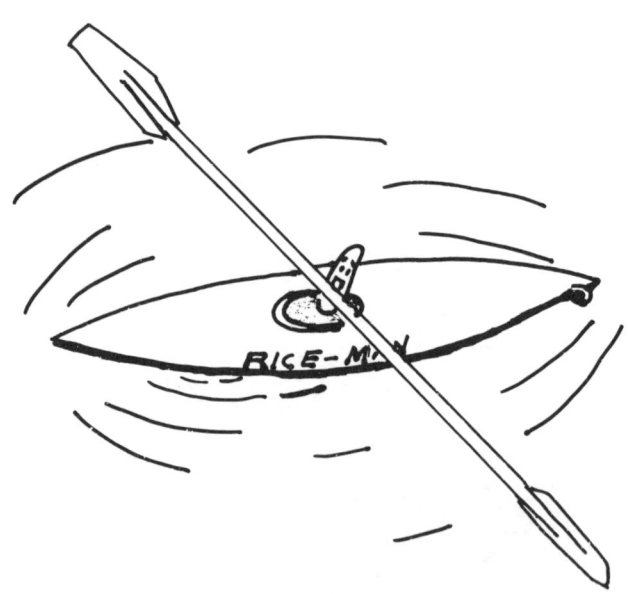

 Cocoa butter is used in the manufacturing of foods, cosmetics, soaps, suntan lotions and pharmaceuticals.

MEAT
(with Nut
Alternatives)

I couldn't leave this loaf alone!

⅔ cup	butter OR margarine	150 mL
¼ cup	brown sugar	50 mL
2	eggs	2
¼ cup	milk	50 mL
1 tsp.	vanilla	5 mL
½ cup	finely chopped chocolate chips	125 mL
1 cup	flour	250 mL
1½ tsp.	baking powder	7 mL
½ tsp.	salt	2 mL
1½ cups	chopped pistachios	375 mL

Cream together butter and brown sugar. Add eggs, milk, vanilla and chocolate chips. Mix well. Stir in flour, baking powder, salt and 1 cup (250 mL) of pistachios. Mix again. Spoon batter into greased 5" x 9" (13 cm x 23 cm) loaf pan and sprinkle with remaining pistachios. Bake at 350°F (180°C) for 40 to 45 minutes. Makes 1 loaf.

 All American and Soviet spaceflights most certainly have had chocolate aboard.

CHOCOLATE PECAN MUFFINS

These are great as an after-school snack!

½ cup	oat bran	125 mL
½ cup	granola cereal	125 mL
¾ cup	milk	175 mL
1	egg	1
½ cup	oil	125 mL
1 cup	flour	250 mL
2 tsp.	baking powder	10 mL
½ tsp.	salt	2 mL
¾ cup	sugar	175 mL
¼ cup	cocoa	50 mL
½ cup	chopped pecans	125 mL
1 tsp.	instant coffee granules	5 mL

Combine oat bran, granola and milk. Let stand until milk is almost completely absorbed. Add egg and oil. Mix well. In a separate bowl, combine flour, baking powder, salt, sugar and cocoa. Stir in pecans and coffee. Add dry ingredients to moist. Stir slightly. Spoon into greased muffin tins and bake at 400°F (200°C) for 20 minutes. Makes 12 medium-sized muffins.

 Yes! Chocolate is the food of love — modern analysis of chocolate reveals small amounts of phenylethylamine, a chemical naturally produced in the brain that increases when people fall in love.

OLD-FASHIONED BROWNIES

Yes, another personal favorite!

⅓ cup	butter OR margarine	75 mL
1 cup	sugar	250 mL
2	eggs	2
1 tsp.	vanilla	5 mL
½ tsp.	salt	2 mL
¾ cup	cocoa	175 mL
⅓ cup	oil	75 mL
½ cup	flour	125 mL
½ tsp.	baking powder	2 mL
1 cup	chopped pecans	250 mL

Melt butter then add sugar and mix well. Stir in eggs, vanilla and salt. In a small bowl combine cocoa and oil. Blend well. Stir chocolate into egg mixture and blend again. Gently stir in flour, baking powder and pecans. Pour batter into greased 8″ (20 cm) square pan and bake at 350°F (180°C) for 30 minutes. Frost with Chocolate Buttercream Frosting, page 11.

 Cocoa butter will never, ever go rancid.

 "TIGER" COOKIES

They're grrreat!

1 cup	chocolate-orange chips	250 mL
1 cup	butter OR margarine	250 mL
1 cup	brown sugar	250 mL
2	eggs	2
1 tsp.	vanilla	5 mL
2 cups	flour	500 mL
1 tsp.	baking soda	5 mL
¼ tsp.	salt	1 mL

Melt chocolate-orange chips over hot water or on MEDIUM (50%) power for 3 minutes. Stir to melt. Cream together butter and brown sugar. Beat in eggs and vanilla. Add flour, baking soda and salt. Blend well. Add melted chocolate and stir just to marble the mixture. Drop batter by spoonfuls onto greased baking sheet. Bake at 375°F (190°C) for 15 minutes. Makes 36 cookies.

Chocolate is composed of over 300 identified components; therefore, it cannot be synthesized.

CHUNKY NUT COOKIES

The crème de la crème de la cookie!

½ cup	butter OR margarine	125 mL
⅓ cup	white sugar	75 mL
¼ cup	brown sugar	50 mL
1	egg	1
1 tsp.	vanilla	5 mL
1⅓ cups	flour	325 mL
½ tsp.	baking soda	2 mL
½ tsp.	salt	2 mL
3 oz.	semisweet chocolate, coarsely chopped* (3 squares)	85 g
½ cup	coarsely chopped almonds	125 mL
½ cup	coarsely chopped pecans	125 mL

Combine butter and sugars. Beat until fluffy. Add egg and vanilla. Mix again. Stir in flour, baking soda and salt. Fold in chocolate, almonds and pecans. Drop batter by spoonfuls onto greased baking sheet. Bake at 375°F (190°C) for 15 minutes. Makes 20 cookies.

* Belgian, Dutch and Swiss chocolate are excellent choices. They may be purchased at specialty chocolate shops.

See photograph on back cover.

CHOCOLATE ALMOND DELIGHTS

Simply delightful! Delightfully simple!

⅓ cup	oil	75 mL
2 oz.	unsweetened chocolate (2 squares)	55 g
¾ cup	water	175 mL
1 cup	sugar	250 mL
1	egg	1
1¼ cups	flour	300 mL
½ tsp.	salt	2 mL
½ tsp.	baking soda	2 mL
1 tsp.	almond extract	5 mL
1 cup	chocolate chips	250 mL
⅓ cup	slivered almonds	75 mL

Heat oil and chocolate in an 8″ (20 cm) square pan at 350°F (180°C) for 4 minutes. Add water, sugar, egg, flour, salt, baking soda and almond extract. Beat until smooth. Spread batter evenly and sprinkle with chocolate chips and almonds. Bake at 350°F (180°C) for 40 minutes.

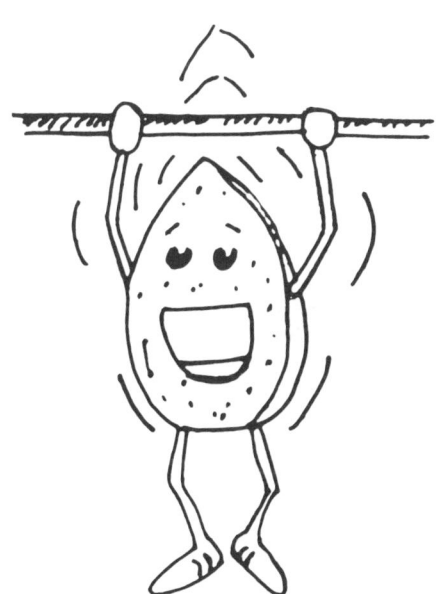

 It has been scientifically proven that chocolate neither causes nor aggravates acne.

CHOCOLATE PEANUT CHEWS

These chewy treats are a snap to make.

2 cups	sugar	500 mL
½ cup	cocoa	125 mL
½ cup	milk	125 mL
¼ cup	butter OR margarine	50 mL
½ cup	peanut butter	125 mL
1 tsp.	vanilla	5 mL
3 cups	oatmeal	750 mL
1 cup	chopped peanuts	250 mL
	whole peanuts for garnish	

In a saucepan combine sugar and cocoa. Stir in milk and butter. Bring to a boil, over medium heat, stirring constantly. Boil gently for 2 minutes. Remove from heat and add peanut butter and vanilla. Stir until smoothly combined. Stir in oatmeal and chopped peanuts. Spread evenly in greased 8″ (20 cm) square pan. Cut into squares and top each with a peanut. Cool until firm.

 Musicians Billy Idol and Prince will not perform on tour unless they have had a feed of chocolate chip cookies.

CHOCOLATE PECAN BARS

Cinnamon really perks up the flavor.

1½ cups	flour	375 mL
½ tsp.	cinnamon	2 mL
¼ tsp.	baking powder	1 mL
½ cup	brown sugar	125 mL
¾ cup	butter OR margarine	175 mL
3	eggs	3
½ cup	brown sugar	125 mL
1 cup	corn syrup	250 mL
¼ cup	flour	50 mL
⅓ cup	chocolate chips, melted and cooled	75 mL
1 tsp.	vanilla	5 mL
¼ tsp.	salt	1 mL
1½ cups	coarsely chopped pecans	375 mL

Combine 1½ cups (375 mL) flour, cinnamon and baking powder. Stir in ½ cup (125 mL) brown sugar. Cut in butter until mixture is crumbly. Press into greased 9″ x 13″ (23 cm x 33 cm) pan. Bake at 350°F (180°C) for 10 minutes. In a separate bowl, combine eggs, ½ cup (125 mL) brown sugar, syrup, ¼ cup (50 mL) flour, cooled melted chocolate, vanilla and salt. Beat until thoroughly blended. Pour onto baked layer. Sprinkle with pecans. Bake at 350°F (180°C) for 30 minutes. Cool and cut into bars.

 "TURTLE" CAKE

Turns a simple cake mix into a rich and chewy delight!

19 oz.	pkg. chocolate cake mix	**520 g**
14 oz.	pkg. caramels	**397 g**
¾ cup	butter OR margarine	**175 mL**
¼ cup	evaporated milk	**50 mL**
1 cup	chopped pecans	**250 mL**
1 cup	chocolate chips	**250 mL**

Prepare cake mix as directed. Pour half of the batter into a greased 9″ x 13″ (23 cm x 33 cm) pan. Bake 15 minutes at 350°F (180°C). In top of double boiler, combine the caramels, butter and milk and cook until smooth. Pour caramel mixture over hot cake in pan. Sprinkle pecans and chocolate chips evenly over the caramel. Pour remaining batter on top. Bake for 25 minutes at 350°F (180°C). Frost with Chocolate Buttercream Frosting, page 11, and serve cake with ice cream, if desired.

 Musician Eddie Van Halen loves M & M chocolate candies but he will not eat the brown ones.

Chocolate Coconut Cream Pie, page 113.
Black Forest Bundt Cake, page 42.

CHOCOLATE ALMOND CAKE

A fabulous cake for any occasion!

¾ cup	butter OR margarine	175 mL
1½ cups	sugar	375 mL
2	eggs	2
2 tsp.	vanilla	10 mL
¾ cup	sour cream	175 mL
2 cups	flour	500 mL
⅔ cup	cocoa	150 mL
½ tsp.	salt	2 mL
2 tsp.	baking soda	10 mL
1 cup	buttermilk	250 mL
1 cup	coarsely chopped, toasted almonds	250 mL

Cream butter and sugar. Add eggs and vanilla and beat well. Blend in sour cream. In a separate bowl, combine flour, cocoa, salt and baking soda. Add flour mixture, alternately with buttermilk, to creamed mixture, then add almonds. Beat well. Pour batter into greased tube pan. Bake at 350°F (180°C) for 60 minutes. Cool cake for 20 minutes before inverting pan to remove cake. Top with Chocolate Glaze, page 64, in which 2 tbsp. (30 mL) of Amaretto or 1 tsp. (5 mL) of almond extract have been substituted for rum or serve cake with Amaretto-flavored whipped cream and sprinkle with toasted slivered almonds.

 CHOCOLATE COCONUT CAKE

The "surprise" center is deliciously delightful!

3 oz.	**unsweetened chocolate (3 squares)**	85 g
½ cup	**butter OR margarine**	125 mL
2 cups	**brown sugar**	500 mL
2	**eggs + 1 egg yolk**	2
1½ tsp.	**vanilla**	7 mL
1 cup	**sour cream**	250 mL
2 tsp.	**baking soda**	10 mL
½ tsp.	**salt**	2 mL
1¾ cups	**flour**	425 mL
1 cup	**boiling water**	250 mL
1	**egg white**	1
⅓ cup	**sugar**	75 mL
1 cup	**coconut**	250 mL
1 tsp.	**flour**	5 mL
	toasted, shredded coconut for garnish	

In a saucepan, melt chocolate over low heat. In a bowl, combine butter and brown sugar, beating until light and fluffy. Add eggs, vanilla, melted chocolate and sour cream. Stir in baking soda and salt. Beat with electric mixer on high speed for 5 minutes. Add flour and beat on low speed until smooth. Pour in boiling water and stir with a spoon until blended. Prepare filling by beating egg white until frothy, then gradually beat in sugar until stiff peaks form. Fold in coconut and flour. Spoon half the batter into greased tube or bundt pan. Spoon coconut mixture over top. Cover with remaining batter. Bake at 350°F (180°C) for 45 minutes. Allow cake to cool for 20 minutes before inverting pan to remove cake. Frost with Fudge Frosting, page 68, and top with toasted, shredded coconut.

CHOCOLATE HAZELNUT CAKE

A very light cake — it can be whipped up at a moment's notice.

4	eggs	4
¾ cup	sugar	175 mL
2 tbsp.	cocoa	30 mL
2 tbsp.	flour	30 mL
2½ tsp.	baking powder	12 mL
1 cup	whole hazelnuts	250 mL

In a food processor, whirl eggs and sugar until thoroughly mixed. Add cocoa, flour, baking powder and hazelnuts (nuts will be chopped in food processor). Blend well. Pour into 2 greased 8'' (20 cm) layer pans. Bake at 350°F (180°C) for 25 minutes. Cool cake layers. Fill and frost with Mocha Frosting, recipe follows.

 MOCHA FROSTING

4 tbsp.	butter OR margarine	60 mL
2 cups	icing sugar	500 mL
2 tbsp.	hot, strong coffee	30 mL
2 tbsp.	cocoa	30 mL
1 tsp.	vanilla	5 mL

Cream butter and icing sugar. Add coffee, cocoa and vanilla. Beat until smooth.

CHOCOLATE PISTACHIO TORTE

A triple-layer treat.

½ cup	cocoa	125 mL
½ cup	boiling water	125 mL
1¾ cups	flour	425 mL
1 tsp.	baking powder	5 mL
1 tsp.	baking soda	5 mL
⅛ tsp.	salt	0.5 mL
½ cup	butter OR margarine	125 mL
2 cups	sugar	500 mL
2	eggs	2
1 tsp.	vanilla	5 mL
1⅓ cups	buttermilk	325 mL
1½ cups	chopped pistachios	375 mL
1 cup	whipping cream, whipped	250 mL
	chopped pistachios for garnish	

Combine cocoa and boiling water. Set aside to cool. Sift together flour, baking powder, baking soda and salt. In another bowl, cream butter and sugar until fluffy. Add eggs, vanilla and cocoa mixture. Blend thoroughly. Add buttermilk and flour mixture, alternately, to creamed mixture. Gently fold in pistachios. Divide batter into 3 greased 8" (20 cm) round layer pans and bake at 350°F (180°C) for 30 to 35 minutes. Remove cakes from oven and cool 10 minutes before removing from pans. To assemble cake, place 1 layer on a cake plate and spread with half the whipped cream. Place second layer on top and spread with remaining cream. Top with remaining layer and frost top and sides with Chocolate Buttercream Frosting, page 11. Garnish with chopped pistachio nuts.

▲ CHOCOLATE "MOOSE" ▲

Light, smooth and scrumptious.

1 tbsp.	unflavored gelatin (7 g envelope)	15 mL
½ cup	sugar	125 mL
¼ tsp.	salt	1 mL
1¼ cups	milk	300 mL
1 cup	chocolate chips	250 mL
2 tbsp.	Irish Cream liqueur	30 mL
1 cup	whipping cream	250 mL
	shaved chocolate for garnish	

In a saucepan, combine gelatin, sugar and salt. Stir in milk and chocolate chips. Stir constantly, over medium heat, until sugar dissolves and chocolate is completely melted. Remove from heat and beat thoroughly. Add Irish Cream liqueur. Chill. Beat whipping cream until stiff. Fold whipped cream into chocolate mixture. Turn into 9" (23 cm) springform pan and chill at least 3 hours. Before serving, garnish with shaved chocolate. Serves 6.

FUDGE NUT PUDDING

A creamy, rich baked pudding that makes its own sauce.

¼ cup	butter OR margarine	50 mL
¾ cup	sugar	175 mL
1½ cups	flour	375 mL
2½ tsp.	baking powder	12 mL
½ tsp.	salt	2 mL
¾ cup	milk	175 mL
¾ cup	chopped walnuts	175 mL
4 tbsp.	cocoa	60 mL
¾ cup	brown sugar	175 mL
¼ tsp.	salt	1 mL
1 cup	boiling water	250 mL

Cream butter and sugar until fluffy. In a separate bowl, combine flour, baking powder and salt. Add flour mixture, alternately with milk, to creamed mixture. Stir in walnuts. Spoon batter into greased 9'' (23 cm) square cake pan. In another bowl, combine cocoa, brown sugar and salt. Sprinkle cocoa mixture over batter. Gently pour boiling water on top. Bake at 375°F (190°C) for 25 minutes. Serve warm, topped with ice cream. Serves 6 to 8.

 A technological institute in the United States has discovered a protein component in cocoa which actually inhibits the growth of bacteria causing tooth decay.

 "ZEBRA" PUDDING

Rich and creamy!

2 cups	milk	500 mL
3 tbsp.	cornstarch	45 mL
¾ cup	sugar	175 mL
4 tbsp.	cocoa	60 mL
1 tsp.	vanilla	5 mL
1	egg	1
2 tbsp.	milk	30 mL
1 cup	whipping cream, whipped	250 mL
6	whole strawberries (optional)	6

In a double boiler, heat 2 cups (500 mL) milk. In a small bowl, combine cornstarch, sugar, cocoa, vanilla, egg and 2 tbsp. (30 mL) milk. Pour cocoa mixture into hot milk, stirring constantly until thick. Cool. Whip cream. Spoon layers of pudding and whipped cream, alternately, into 6 parfait glasses. Chill. Garnish with whole strawberries, if desired. Serves 6.

Chocolate is a natural food made from natural ingredients. The body needs a variety of nutrients to stay healthy. Chocolate, when combined with other natural products such as sugar, milk, fruit, vegetables, cereals and nuts, contributes to maintaining dietary balance. It contains protein, fat, carbohydrates, calcium, phosphorus, iron, potassium, vitamin A, thiamine, riboflavin and niacine.

 FUDGEY PECAN PIE

They'll be asking for seconds.

1	unbaked 10″ (25 cm) pie shell	1
½ cup	sugar	125 mL
½ cup	flour	125 mL
⅓ cup	cocoa	75 mL
¼ tsp.	salt	1 mL
1¼ cups	corn syrup	300 mL
3	eggs	3
⅓ cup	butter OR margarine, melted	75 mL
1½ tsp.	vanilla	7 mL
½ cup	chopped pecans	125 mL
1 cup	pecan halves	250 mL

Prepare pie shell. In a medium-sized bowl, combine sugar, flour, cocoa and salt. Stir in syrup. Add eggs, butter and vanilla. Beat until blended. Stir in chopped pecans. Pour mixture into pie shell. Arrange pecan halves over filling. Bake at 350°F (180°C) for 50 to 55 minutes. Serve with whipped cream.

See photograph on back cover.

Chocolate is a natural flavor enhancer — try chocolate with coffee, mint, cinnamon or vanilla. It is the perfect foil for fruits, spirits and liqueurs.

CHOCOLATE COCONUT CREAM PIE

A final personal favorite!

1⅓ cups	crushed chocolate wafers	325 mL
1½ cups	shredded coconut	375 mL
½ cup	butter OR margarine, melted	125 mL
1 tbsp.	unflavored gelatin (7 g envelope)	15 mL
2 cups	milk	500 mL
1	egg	1
1 tbsp.	cornstarch	15 mL
⅓ cup	sugar	75 mL
½ cup	chocolate chips	125 mL
1 tsp.	vanilla	5 mL
1 cup	whipping cream, whipped	250 mL
	chocolate curls for garnish	

Combine wafers, ½ cup (125 mL) coconut and butter. Press firmly into greased 9″ (23 cm) pie plate. Chill. Sprinkle gelatin over ½ cup (125 mL) of milk and set aside. In a saucepan, heat remaining milk. In a bowl, combine egg, cornstarch and sugar. Pour into hot milk. Cook over medium heat, stirring constantly until mixture thickens and comes to a boil. Remove from heat and add chocolate chips, gelatin mixture, vanilla and remaining coconut. Blend thoroughly. Chill until slightly thickened. Fold in ½ of the whipped cream and pour into the crust. Top with the remaining whipped cream and garnish with chocolate curls. Chill until ready to serve.

See photograph, page 103.

"GRASSHOPPER" PIE

Light, frothy and delicious!

1½ cups	crushed chocolate wafers	375 mL
⅓ cup	butter OR margarine, melted	75 mL
¾ cup	finely chopped hazelnuts	175 mL
1 tbsp.	unflavored gelatin (7 g envelope)	15 mL
¾ cup	sugar	175 mL
2	eggs, separated	2
¾ cup	milk	175 mL
¼ cup	green crème de menthe	50 mL
2 tbsp.	white crème de cacao	30 mL
1½ cups	whipping cream, whipped	375 mL
	chocolate curls for garnish	

Combine wafers, butter and hazelnuts. Press into bottom and up the sides of a 10" (25 cm) pie plate. In a saucepan, combine gelatin and sugar. Beat together egg yolks and milk. Gradually add to gelatin mixture. Cook over low heat, stirring constantly, until gelatin and sugar are dissolved. Remove from heat. Stir in crème de menthe and crème de cacao. Chill. Beat egg whites until frothy. Fold egg whites and 1 cup (250 mL) of whipping cream, whipped, into chilled gelatin mixture. Turn into crust. Chill. Garnish with ½ cup (125 mL) whipping cream, whipped, and chocolate curls.

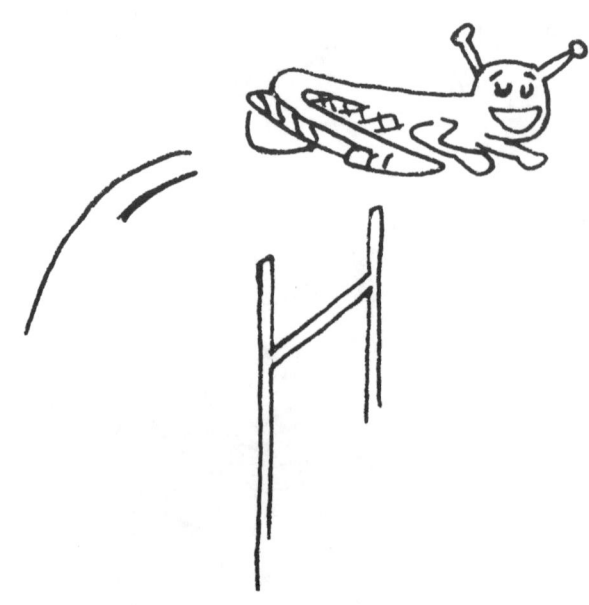

NUTTY CHOCOLATE FUDGE

This fudge has never failed me!

½ cup	butter OR margarine	125 mL
¾ cup	cocoa	175 mL
2 cups	sugar	500 mL
⅛ tsp.	salt	0.5 mL
¾ cup	evaporated milk	175 mL
2 cups	miniature marshmallows	500 mL
1 tsp.	vanilla	5 mL
1 cup	chopped walnuts OR pecans	250 mL

In a saucepan, melt together butter and cocoa, stirring until smooth. Add sugar, salt and evaporated milk. Bring mixture to a boil over medium heat. Boil and stir constantly for 5 minutes. Remove from heat and add marshmallows and vanilla. Stir until marshmallows have melted completely. Cool 20 minutes, then beat until mixture loses its gloss. Fold in nuts. Spread in a greased 9" (23 cm) square pan. Refrigerate until firm.

INDEX

INDULGE A FRIEND

Please send me _____ copies of **"CHOCOLATE . . . A HEALTHY NEW IMAGE!"**
at $9.95 per book plus $1.50 (total order) for shipping and handling.

Number of books _____ x $9.95 = _____

Postage and handling _____ = $ $1.50

Total enclosed _____ = _____

U.S. and international orders payable in U.S. funds

NAME _____

STREET _____

CITY _____ PROV./STATE _____

COUNTRY _____ POSTAL CODE/ZIP _____

Please make cheque or money order payable to
   "Sweet Dreams Publishing"
   P.O. Box 3856
   Regina, Saskatchewan
   S4P 3R8

For fund-raising or volume purchases, contact SWEET DREAMS PUBLISHING for volume
rates. (306) 545-0751. Please allow 3-4 weeks for delivery.

Price is subject to change.

INDULGE A FRIEND

Please send me _____ copies of **"CHOCOLATE . . . A HEALTHY NEW IMAGE!"**
at $9.95 per book plus $1.50 (total order) for shipping and handling.

Number of books _____ x $9.95 = _____

Postage and handling _____ = $ 1.50

Total enclosed _____ = _____

U.S. and international orders payable in U.S. funds

NAME _____

STREET _____

CITY _____ PROV./STATE _____

COUNTRY _____ POSTAL CODE/ZIP _____

Please make cheque or money order payable to
   "Sweet Dreams Publishing"
   P.O. Box 3856
   Regina, Saskatchewan
   S4P 3R8

For fund-raising or volume purchases, contact SWEET DREAMS PUBLISHING for volume
rates. (306) 545-0751. Please allow 3-4 weeks for delivery.

Price is subject to change.